T. Dandekar (Ed.) · *RNA Motifs and Regulatory Elements*

Springer

Berlin
Heidelberg
New York
Barcelona
Hong Kong
London
Milan
Paris
Tokyo

Thomas Dandekar (Ed.)

RNA Motifs
and Regulatory Elements

Second Edition

With Contributions by
Peter Bengert, Thomas Dandekar, Dirk Ostareck,
and Antje Ostareck-Lederer

With 39 Figures and 10 Tables

Springer

Editor

Prof. Dr. THOMAS DANDEKAR
Europäisches Laboratorium
für Molekularbiologie (EMBL)
Meyerhofstr. 1
69012 Heidelberg, Germany
and

Universität Freiburg
Institut für Molekulare Medizin
und Zellforschung
Breisacherstr. 66
79106 Freiburg, Germany

Contributors

Dr. PETER BENGERT
Universität Heidelberg
Biochemie-Zentrum (BZH)
69120 Heidelberg, Germany
and
Universität Freiburg
Institut für Molekulare Medizin
und Zellforschung
Breisacherstr. 66
79106 Freiburg, Germany

Prof. Dr. THOMAS DANDEKAR (*see above*)

Dr. DIRK OSTARECK
Dr. ANTJE OSTARECK-LEDERER
Europäisches Laboratorium
für Molekularbiologie (EMBL)
Meyerhofstr. 1
69012 Heidelberg, Germany

Cover illustration: Left and background: Internal Ribosome Entry Site in the RNA genome of the Cricket Paralysis Virus (Courtesy of Peter Sarnow, Dept. of Microbiology and Immunology, Stanford Univ. School of Medicine, Stanford, CA, USA). Right and front: Hairpin ribozyme crystal structure, bound U1A protein is shown at the bottom (Courtesy of Adrian R. Ferré-D'Amaré, Div. of Basic Sci., Fred Hutchinson Cancer Res. Center, Seattle, WA, USA).

ISBN 3-540-41701-X Springer-Verlag Berlin Heidelberg New York

First edition 1998 "Regulatory RNA" by T. Dandekar and K. Sharma, ISBN 3-540-64343-5

Library of Congress Cataloging-in-Publication Data

RNA motifs and regulatory elements / T. Dandekar (ed.); contributors, P. Bengert ... [et al.].--2nd ed.
 p. cm
 Rev ed. of: Regulatory RNA / Thomas Dandekar. 1998.
 Includes bibliographical references and index.
 ISBN 354041701X (hardcover : alk. paper)
 1. RNA. 2. Messenger RNA. 3. Genetic regulation. I. Dandekar, Thomas, 1960- II.
Bengert, P. (Peter) III. Dandekar, Thomas, 1960-Regulatory RNA.
QP623 .D36 2001

Springer-Verlag Berlin Heidelberg New York a member of BertelsmannSpringer Science+Business Media GmbH
http://www.springer.de

© Springer-Verlag Berlin Heidelberg 2002
Printed in Germany

The use of general descriptive names, registered names, trademarks, etc. in this publication does not imply, even in the absence of a specific statement, that such names are exempt from the relevant protective laws and regulations and therefore free for general use.

Cover design: D&P, Heidelberg
Typesetting: Camera-ready by the editor
SPIN 10768561   31 /3130 - 5 4 3 2 1 0 - Printed on acid-free paper

Dedicated to our mothers, Monika and Helga,
who gave us more than life.
    A.O.-L. and D.H.O.

Dedicated to my mom Doris, my dad Clemens
and Julia for their support.
    P.B.

To my Readers, my Nice advisors
And my wife Gudrun.
    T.D.

Preface

We earlier tackled the topic RNA motifs and regulatory elements with a book called Regulatory RNA (Dandekar and Sharma, 1998). Our new book not only reflects a shift in emphasis further to the motifs and their implications, but was also necessary to keep up with the pace of this rapidly expanding field. As in the previous edition, we have made an effort to cover the data extensively and include latest results. Nevertheless, due to space limitations, we cannot avoid being short in some places and incomplete (due to time limitations in preparing this book) in others. We apologize therefore to our coworkers in the field whose work was not incorporated or has some other unnoticed shortcomings in our representation, and will be happy to incorporate any comments and criticism in the next edition.

Again, this book would not have been possible without the numerous colleagues at EMBL and the universities of Heidelberg and Freiburg who have critically read the manuscript and made valuable suggestions.

This book should further stimulate research in the exciting field of regulatory RNA motifs and RNA elements, an endeavour which has just begun, compared to the numerous instances in the living cell still waiting to be discovered.

Contents

1. Introduction

1.1 Book Overview

After the general introduction, RNA motifs and their functional implications are explained with different specific tables and examples in Chapter 2. Experimental (Chap. 3) and theoretical (Chap. 4) standard techniques to identify RNA have been kept, as in the pre-runner of this book, as these techniques in the field remain the most powerful tool to identify new instances of regulatory RNA motifs. However, we suspect that these may change in the future with large-scale array techniques in functional genomics becoming more and more powerful. In a similar way, network models of the cell are also becoming increasingly popular and may soon provide solid tools of their own to identify new regulatory RNA networks. Dynamics as well as medical implications of RNA motifs (Chap. 5) have experienced interesting new contributions during the past 3 years and become increasingly more important. Similarly we note that unexpected new hot topics (Chap. 6) of research in the field have come up, such as translational silencing and IRES mediated translation initiation.

The interested student is heartily welcomed to this book; however, probably he or she will be on an advanced level, for example doing a thesis in the field. For the specialist the book should be a well-suited reference, summary and incentive with specific chapters covering different angles of interest on regulatory RNA elements.

1.2 RNA, the Underestimated Molecule

RNA has long been underestimated in its capacities, originally considered an uninteresting contaminant of the more important proteins. Torbjörn Caspersson was among the first during the late 1930s to suggest from microscopic studies that RNA has an important role in the cytosol of the eukaryotic cell. Using chemical approaches, Jean Brachet (reviewed in Brachet 1987) confirmed Caspersson's findings and found in cytosolic ribonucleoprotein particles a first hint of the ribosome. Only in the 1950s did radiolabelling studies show that amino acids were assembled sequentially on ribosomes to form proteins. In 1955 Crick formulated the adaptor hypothesis, in which translation was thought to occur through the mediation of transfer RNA (tRNA) adaptor molecules. At about this time Paul

Zamecnik and Mahlon Hoagland started to discover tRNA: during protein synthesis C^{14}-labelled amino acids became transiently bound to a low molecular mass fraction of RNA, soluble RNA or sRNA (Holley et al. 1965). In 1958, including his adaptor hypothesis, Francis Crick formulated the central dogma of molecular biology, the flux of information from DNA via RNA to proteins (reviewed in Crick 1970).

Sidney Brenner and coworkers (1960) were the first to describe mRNA as "an unstable intermediate carrying information from genes to ribosomes for protein synthesis". In 1961 Jacob and Monod published their seminal paper on the existence of mRNA and operons and explained how the transcription of operons was regulated. The involvement of RNA in the synthesis of proteins was established in 1963 by Watson. Hall and Spiegelman in 1964 showed sequence complementarity of T2-DNA and T2-specific RNA. As a result of these studies, the exciting discovery of the genetic code began (Crick 1965). This culminated in the RNA being established as the messenger, the transient carrier of genetic information. As a further surprise, in 1975 Thomas Cech and colleagues (reviewed in Cech 1993) showed that RNA could also be catalytically active in addition to carrying genetic information. In view of these additional capabilities, RNA is now generally considered to have been critical for the evolution of life before the genetic code had evolved (reviewed in Gesteland et al. 1999): RNA could thus have performed dual functions in ancestral cells for which DNA and proteins are now used. DNA stores more stable information than RNA and proteins can be more catalytically active than RNA, but only RNA can do both. Reverse transcriptase found in retroviridae (Temin 1972), explained the emergence of DNA as a more solid permanent form of storage but challenged the "central dogma of molecular biology".

Increasing evidence supports the RNA world. Splicing, the metabolism of precursor mRNA involving the excision of internal sequences is another surprise in this respect. Initial studies (by the groups from Phillip Sharp and Richard Roberts; Chow et al. 1977, Berget et al. 1977) compared mRNA and DNA from adenovirus, and found viral DNA segments which were not contained in the mature mRNA. Subsequently such gene interruptions were found in ovalbumin and ß-globin genes (Crick 1979; Chambon 1981; Perry 1981). The RNA world theory is further supported by the observation of many nucleotide cofactors such as NADH, FAD found in basic metabolic enzymes which may be considered as a vestige from the transition of an RNA-dominated world to a protein-dominated one. Moreover, many new different types of ribozymes are now known which catalyze a large variety of reactions. Catalytic ribonucleoprotein particles in which the active RNA molecule is the principal part are another supporting observation such as RNAse P (Altman et al. 1993; Warnecke et al. 1996; Gesteland et al. 1999). The double capacity of the RNA molecule, both to carry information and to be catalytically active, is the foundation for its many functions and uses in the living cell.

1.3   Examining Regulatory RNA Elements

Information storage and transport are required for many tasks of the cell. RNA is an optimal molecule for this. Moreover, RNA metabolism creates many different levels where information may be stored, transported or specifically released in the cell. For such regulatory steps and other specific functions, specific structures in the RNA, so-called RNA elements or motifs, have evolved (several ones in RNA are shown in Fig. 1.1; see below for details). They allow specific and controlled release of information in the cell from an RNA containing such an element (Dandekar and Hentze, 1995). An RNA element (sometimes also called RNA motif, regulatory element, binding site or simply RNA signal, each term stressing different aspects of the function in which the RNA segment is critically involved) is a small segment of an RNA molecule which is essentially required for a specific interaction of the RNA with the enviroment. This includes interaction with itself (autocatalytic cleavages, for instance) or other binding partners (RNAs and RNA "recognizing", i.e. binding proteins). Essential features of the RNA segment are specific nucleotides and the RNA structure. RNA structure is formed by base-pairing of different parts of the RNA molecule to each other. There are two possibilities: direct base pairing forms an RNA structure, this is called secondary structure. If there are higher-level interactions, this is called tertiary structure. Examples for tertiary structure are interactions between two secondary structures, or between a secondary structure and an additional single-stranded region. Tertiary interactions often happen between nucleotides separated by long (100 or more base pairs) distances, however, shorter tertiary interactions are also known (see Fig. 1.2, right).

As an introductory RNA illustrating the combination of sequence, secondary structure and further structural features, the rev-response core element (RRE) from human immuno deficiency virus (HIV) is shown in Fig. 1.2. The nucleotide sequence of the core motif where rev protein binds the RNA is shown in full. Besides the primary sequence, base pairing between both strands is apparent, forming a stem. Further features of the secondary structure are nucleotides not paired but "bulged out", the A and U on the right side of the stem. The motif itself is part of a more complex structure. The whole rev RNA structure encompasses 234 nucleotides and some of the additional secondary structure, composed of RNA stems and single-stranded RNA loops are sketched. Opposing base-paired RNA strands form stems, shown as parallel lines; unpaired single-stranded RNA forms "loops", shown as open circles (the complete element includes further sequences close to the core consensus structure shown).

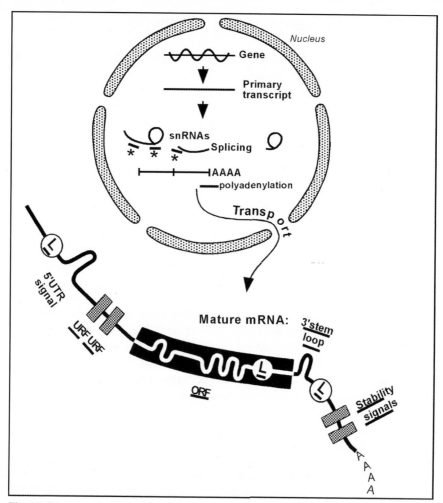

Fig. 1.1. Regulatory motifs involved in messenger RNA metabolism.
The figure summarizes regulatory motifs (indicated by *short solid lines*). *Top* Regulatory motifs involved in processing the mRNA precursor. After the gene is transcribed the primary transcript undergoes splicing, which requires several RNA-RNA interactions and recognition of RNA motifs. Schematically indicated by * in the nucleus interactions between snRNA and mRNA precursor at the 5' splice site, 3' splice site and the branch point are shown; not shown are numerous motifs present in the snRNAs involved in the splicing process. After splicing the exons are assembled and polyadenylated (AAAA). This process as well as the subsequent transport from the nucleus into the cytoplasm again involves different RNA motifs on the mRNA. Motifs in the mature mRNA (*bottom part*) include: localization signals (*L*) in the 5'UTR and 3'UTR; stem loop structures in 5' UTR, open reading frame and 3' UTR; short upstream reading frames (uORF) in the 5' UTR; regulatory signals in the open reading frame encoded in its secondary structure and primary sequence exploiting the degeneracy of the genetic code and involving also localization signals; regulatory signals in the 3' UTR include stem-loop structures, further localization signals, sequences involved in mRNA stability and poly(A) tail (AAAA) formation

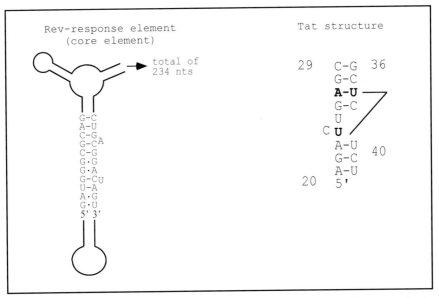

Fig. 1.2. Schematic diagramm depicting RNA motifs.HIV rev serves to illustrate a simple example. The core motif where rev protein binds the RNA is shown in full. Base-pairing between the 5' and 3' strands leads to the formation of a stem. Double stranded regions are drawn as *parallel lines* and unpaired single-stranded regions are drawn as *loops*. Watson Crick base pairs between A and U as well as between C and G are shown as -; all other hydrogen bonding interactions between different nucleotides are shown as throughout the book. A specific feature of the secondary structure are nucleotides (the A und U on the right side of the stem) which are not part of the base-paired stem but are " bulged out" of this RNA stem structure.The HIV rev core motif itself is part of a more complex structure, the complete rev element, which encompasses 234 nucleotides. The complete structure also forms a complex tertiary structure by long distance interactions of the RNA.

The Tat structure on the right shows a stem, which is formed by base pairing between RNA strands. It illustrates a very simple tertiary interaction, in which three nucleotides,shown in bold are connected because a uracil contacts an A-U Watson-Crick base pair. Further details on the particular motifs shown can be found in Malim and Cullem (1994).

Besides this secondary structure, the complete structure also forms tertiary interactions in which distal parts of the RNA interact with each other. For comparison, the *trans*-activation response element (TAR) from HIV on the right shows another regulatory RNA structure. Again, a stem is formed between base-pairing RNA strands; however, this RNA element also illustrates a very simple tertiary interaction. Apart from a standard Watson-Crick base pair between A and U, the three nucleotides shown in bold are connected by a tertiary interaction indicated by the lines forming a V on the right. Further details on the particular motifs shown are described in Cullen (1994). As evident from the figure and also the case for catalytic RNA structures, the function of the element depends both on critical sequence features and nucleotides as well as specific base pairing and secondary structure. A search for related RNA structures with potential similar function as they have similar binding partners (or, if catalytic, similar substrates) must take into account all the features of primary sequence and its detailed RNA structure. The function in the case of the rev-response element (RRE) is binding of the rev protein by this RNA element (stronger binding requires multiple copies of the RRE), but in vitro one core element is sufficient for affinity selection for rev protein binding. A combination of features in sequence and structure is required by most regulatory RNA elements for specific interactions of RNA, such as protein binding, basepairing to another RNA or modifying a nucleic acid bond. The direct interaction, in this case the rev protein binding, has further consequences, in that the viral RNA is not spliced but is very quickly exported from the nucleus. Similarly, other regulatory effects start from interactions of a simple RNA element or RNA motif with a partner (other parts of the molecule, another RNA or a protein). Regulatory effects include regulation of translation, processing of RNA, catalytic modification of other RNA molecules, transport and position in the cell or the stability of the RNA transcript and expression of the encoded protein. However, in many cases the regulatory interaction has consequences for the whole RNA molecule (e.g. its chemical modification), more general effects on the cell (e.g. cellular iron level) or even on the whole organism (e.g. sex-specific splicing or in development; see tables in Chap. 2).

Regarding the rev response element, experimental examination shows that each of these different levels in the structure are required for the functioning of the element and for recognition by the rev protein. Interestingly, analysis of the rev response element confirms that it is critical for rev-protein binding both in the natural HIV virus and by an artificial selection scheme in SELEX experiments (Türk 1997; see below and Chap. 3). Both studies show that critical nucleotides include the central non canonical base pair G•G. A non-canonical base pair is indicated throughout this book by • and is a hydrogen-bonding interaction but generally less stable than the standard base pairings. The standard RNA base pairings between A and U (two hydrogen bonds) or between C and G (three hydrogen bonds) are indicated by a dash throughout this book. After scrutinizing the important parts of primary sequence and structure, this example motif, the Rev response element, can be applied in database searches to find human RNAs which mimic this viral element by a similar RNA structure (Dandekar and Koch 1996). Chapter 4 describes general techniques for the application of this principle to any RNA motif.

      The ribose in RNA has a free 2'-hydroxyl bond available for chemical reactions in contrast to DNA yielding catalytically active RNA structures in contrast to desoxy ribose in DNA. Segments or parts of an RNA structure required for different steps of a catalytic reaction are known as catalytic RNA motifs such as the hammerhead and the hairpin (see Chap. 2). Reactions mediated by catalytic RNA generally proceed 2-4 orders of magnitude more slowly than in proteins. RNA catalyzed reactions are versatile (Lohse and Szostak 1996) including *trans-*

esterification, RNA excision, ligation or partial extension, specific binding tuned to a broad range of chemical ligands, splicing, *trans*-splicing and even amino acid transfer reactions. As is the case in regulatory RNA elements, the catalytic potential of the RNA relies (or can be predicted, see chap. 4) on the specific nucleotides, secondary and tertiary structural features composing the different catalytic RNA elements.

1.4   Regulation of Gene Expression

Generally, only in areas where regulation by RNA has properties superior to proteins or DNA are regulatory RNA elements expected to play a prominent role in regulation.

Regulating Information Release by mRNA

A first area where regulation by RNA containing specific regulatory elements might be superior comprises those instances where information is transported by RNA and released selectively. This process opens up new and different levels of cellular regulation such as the pathway of messenger RNA, i.e. RNA processing, its transport, translation and decay (Fig. 1.1). Some of the involved RNA elements are shown as small black bars regulating different steps in the pathway of a typical mRNA molecule.

The availability of the information carried in a messenger RNA for recognition by the ribosome and synthesis of different proteins can be regulated by specific RNA elements structures in the mRNA (posttranscriptional regulation): these generally reside in the flanking 5' or 3' untranslated regions so as not to interfere with protein coding in the reading frame. Specific interaction partners such as antisense RNA or specific binding proteins recognize these. Their binding may block translation (often in the 5' UTR, e.g. IRE) or inhibit mRNA degradation (often in the 3' UTR) and may itself be modulated by different ion concentrations (Melefors and Hentze 1993).

Further regulatory elements include another type of regulation first exemplified in the 15-lipoxygenase mRNA 3' UTR (Ostareck-Lederer et al., 1994, Ostareck et al. 1997, 2001) and motifs in long stretches of 3' UTR with unknown exact composition (e.g. estrogen receptor mRNA; Kenealy et al., 1996, 2000; 3' UTRs from mRNA in plasmodia, Ruvolo et al. 1993).

Not only temporal but also spatial patterns can be specifically encoded in RNA structures. Localization signals are encoded in mRNAs synthesized during development such as *oskar* mRNA (Erdélyi et al. 1995). *Trans*-acting factors (often proteins) localize the mRNA to a specific cellular compartment (Breitwieser et al. 1996), mediating subsequent pattern formation in the embryo.

Specific short motifs including methylation in snRNAs allow their export as new synthesized snRNAs from the nucleus and their association with proteins in the cytoplasm with subsequent reimport.

This is only a small selection of the different RNA types and RNA elements. Their richness is far greater. Major new areas keep coming up such as new antisense RNAs and transcriptional silencing as well as regulation of chromatin structure, e.g. in *Drosophila* and in mammalian X chromosomal inactivation by the XIST-RNA as the methylation and imprinting by H19 RNA (see Chap. 6).

1.5   Catalytic Active RNA

Many steps of RNA metabolism involve direct RNA cleavage or RNA processing by catalytic RNA as an essential component. The "RNA world" hypothesis (Gesteland et al., 1999) postulates that catalytic RNA originally performed all catalytic activities in the cell and subsequently more and more functions were taken over by proteins. RNA may serve as a protein scaffold, but often it is involved in direct RNA-RNA interactions. Furthermore, in a growing number of specific cases RNA has proven direct catalytic activity. Such RNA-enzymes are called ribozymes. In many examples RNA is an important part of the complete catalytic complex, though it is not strictly proven that the RNA component alone is catalytic. A first group in this category are self-splicing RNAs, such as the *Tetrahymena* group I intron (Cech 1993) and in mitochondrial pre-mRNA transcripts. Further examples are the hammerhead and related catalytic RNA motifs found in RNA viridae performing different cleavage reactions.

In splicing, the processing required to yield the mature mRNA from a longer precursor containing introns and exons, specific evolutionary conserved sequences in the pre-mRNA molecule define 5' splice site, branch-point and 3' splice site (see Chap. 5 and 6) and led also originally to the discovery of the interacting RNA partners, the snRNAs (Lerner et al., 1980). Interacting factors such as snRNAs have also specific RNA motifs connected to their function. Examples are the Sm site, which is a binding target for interacting common proteins, or the Y motif in the interaction between snRNA U4 and U6 (Li and Brow, 1996).

In parasites such as nematodes and trypanosomes, splicing can also occur in *trans* (Laird 1989; Nilsen 1995): leader RNA sequence is attached to most or some mRNAs in organisms where this reaction occurs. Continuing investigations show that this reaction may be far more widespread than previously thought. Specific RNA motifs which are required for catalytic function in *trans*-splicing RNAs were found in vertebrate mRNA (Dandekar and Sibbald 1990) for this there is also some experimental evidence (Bruzik and Maniatis 1992; Vellard et al. 1992; Chap. 6).

RNAse P is required in particular for tRNA processing in prokaryotes, RNAse MRP for 5.8S rRNA processing in eukaryotes. Both activities rely on an essential RNA component.

Probably there are many more reactions in which the RNA component is more important than previously thought. In processing of the ribosomal RNA transcript, an astounding number of RNA cofactors acting in *trans*, the small nucleolar RNAs, have been identified (Venema and Tollervey 1999). Functional RNA structures and motifs involved in this complex processing reaction are numerous, for instance the guide snoRNAs which contain RNA motifs to direct rRNA methylation (Kiss-Laszlo et al. 1996) as well as for pseudouridylation (Ganot et al. 1997).

New discoveries such as RNA as a principal component in telomerase (Chap. 6) and in the catalytic centre of the ribosome (Ban et al. 2000) indicate that the future will extend this list of examples for important catalytic functions supported by RNA.

1.6 Combination of RNA and Proteins

Ribonucleoproteins are a powerful combination of an RNA containing a regulatory motif, and specific proteins recognizing it, ranging from small nuclear ribonucleoproteins (snRNPs) to the ribosome and the signal recognition particle. In the iron responsive element (IRE) is an RNA element recognized and bound by the iron-responsive element binding protein (IRP) if iron levels are low. These and many further examples of this theme indicate that even more such complexes will be found in the future.

1.7 RNA Motifs and Elements Indicate Conserved Function

In the discussed examples and areas, regulation mediated by an RNA molecule is more effective than conceivable alternatives such as signal cascades or receptor-ligand interactions. In each of these cases the specific function of the RNA is dependent on specific RNA nucleotides and structures. An exact description of these RNA features enables them to be recognized in newly characterized RNAs, either by computational (Chap. 4) or experimental techniques (Chap. 3). This reveals further examples of this RNA motif, the mode of regulation and its connected function. Moreover, phylogenetic comparisons reveal which parts of the RNA may easily change and which other sites have been conserved even in very different organisms, e.g. in U6 snRNA (Li and Brow 1996). If putative interacting regions from two RNAs are compared in many species, one can test for "compensatory" base exchanges: if one of the partner RNAs has changed its nucleotide sequence in one species so that it would disrupt the basepairing to the partner RNA, there is a suitable "compensatory" change in the corresponding partner RNA sequence from that species. The material provided by evolution can be augmented further by direct mutagenesis experiments (Dandekar and Tollervey 1992). These and many similar experiments on the different types of regulatory RNAs underline that the elements contained in RNA are no abstract entities but real domains with a specific function. A modern alternative to reveal their specific requirements are the so-called SELEX (systematic evolution of ligands by exponential enrichment) experiments (Türk 1997). In a standard setting a factor with RNA-binding activity is used to affinity-select RNAs which bind to it from a starting pool of random RNAs. RNAs which bind with an improved affinity over the random level are used to construct a new generation of varied yet related molecules from which a further enhancement of affinity is selected. Molecules obtained from such an approach typically display not the same sequence as found in nature, but this experimentally derived description of the RNA motif highlights the conserved regions and structures essential for the function contained in the RNA motif (see Chap. 3).

1.8   Context Specificity of RNA Within the Cell

RNA represents a unique class of molecule in the cell due to its ability to carry both information and catalytic activity. Both functions depend on interactions of the RNA with other molecules. This stresses the basic but important fact that functional RNA is context-dependent. An instructive example is an artificially synthesized random oligonucleotide injected into a cell and an RNA isolated from the same living cell with identical sequence length: in most of the cases the artificially injected RNA is simply degraded as there is no useful interaction with the components of the cell; the random sequence is not understood as a signal. A philosophical speculation we could make here would stress the context specificity as the critical difference between a "living" and a "dead" molecule as a general criterion for life. Nevertheless, this context specificity can also be exactly measured and explored if for instance yeast transformants complemented by different RNA constructs after a gene disruption are counted and the number of transformants from random mutations is compared with those constructs in which the nucleotides for critical RNA interactions were conserved. A more practical consequence of the context specificity of regulatory RNA motifs is that such interacting regions mutate far less than other regions in the RNA. This can help to observe and identify interacting regions in a newly characterized RNA and even be used as a starting information to search for the correct interacting partner, for instance between snoRNAs and ribosomal RNAs (Kiss-Laszlo et al. 1996). Another example would be U4 mutagenesis and the highly conserved Y-motif, the interaction area with U6 RNA. As snRNAs have many partners, including different protein partners, snRNAs have a highly conserved structure, which is apparent even between mammals and different yeasts (Brow and Guthrie 1988;

Dandekar et al. 1989;). A structure which is conserved in very different organisms points also to a basic function important for cell survival in very different organisms. Both factors, a basic and important function in all organisms with a cell nucleus (eukaryotes) and critical interaction partners, are reasons for the conservation of RNA structure and RNA motifs in snRNAs and in splicing. Furthermore. this principle of basic functions and conserved structure can be generalized. Many newly identified RNAs testify to this, most clearly in those cases where a whole molecule and not only smaller motifs are conserved between distantly related species. Well-known examples are the evolutionary conserved ribosomal RNAs from which many new species are being identified. The suprising multitude of snoRNAs recently discovered and rapidly augmented is a newer example. Conserved structure in this case is partly due to interactions with unrevealed partners (in several rRNA-processing steps) and partly due to understood functional interactions (ribosome methylation guide RNAs; Kiss-Lasló et al 1996)

The context specificity becomes even more prominently visible in RNA localization signals (see above) as here the RNA signal is in fact critically necessary to establish the context in the cell, otherwise the different parts of the cell (or the tissue in later steps of differentiation) miss important organizing cues to establish pattern gradients and body axes (Erdelyi et al. 1995). How important conserved regulatory RNAs in fact are is most strikingly revealed by complementation experiments, for instance the 4.5S *E.coli* RNA by the human 7S RNA, the core RNA of the signal recognition particle (Ribes et al. 1990).

2. Instances of Functional RNA

2.1 Overview

This chapter gives a summary of the different instances of functional RNA. Even while this book was being written, new examples were being described, and thus the list presented is necessarily incomplete. It shows, however, the areas already evident where functional RNA (Table 2.1) and different and specific regulatory motifs (Table 2.2) are important. Revealing and describing the full variety of regulatory RNA is an important prerequisite of future research efforts aiming to understand the network of gene expression on the transcriptional and posttranscriptional level. To provide an incentive to look for new RNA structures, after describing the different RNA motifs and families, the next two chapters explain helpful tools to reveal new regulatory RNAs and regulatory RNA motifs.

RNA motifs which are important for the processing and function of RNAs are based on the combination of secondary and tertiary structure elements, formed both, intra– and intermolecular; examples are shown in Fig. 2.1. Important RNA motifs are already introduced and shown in the figures of this chapter, but details of their function and dynamics are discussed in Chapters 5 and 6 in the context of a more general review of regulatory RNAs.

In many cases, only one or a few RNA structures were initially known. For instance, in the case of iron-responsive element (IRE), originally only the IRE in the 5' UTR of ferritin mRNA (Hentze et al. 1987) and five IREs in the 3' UTR of the transferrin receptor mRNA (Müllner and Kühn 1988) had been characterized. However, experimental characterization (Chap. 3) and computer-aided searches and search programs (Chap. 4) can subsequently reveal a whole family of mRNAs, which utilize a similar regulatory structure and thus follow a similar functional principle. IRE-bearing mRNAs, involved in the control of cellular iron metabolism, are translationally regulated *via* the IRE-binding protein (IRP), which interacts with the IRE depending on the cellular iron status.

In other cases the function of regulatory RNA elements remains to be elucidated. For instance, *trans*-splicing RNAs were originally known only in trypanosomes (Laird 1989). Subsequently computer-aided search showed that *trans*-splicing is widespread and identified putative candidates in vertebrates (Dandekar and Sibbald 1990). *Trans*-splicing has subsequently been shown experimentally to occur in HeLa cells, if a suitable leader sequence is provided, (Bruzik and

Maniatis 1992) and several exciting new instances of *trans*-splicing in different organisms have been described in the meantime (Eul et al. 1996; Shimizu et al. 1996). The question, which of the other candidate RNAs actually carry out *trans*-splicing, is still open.

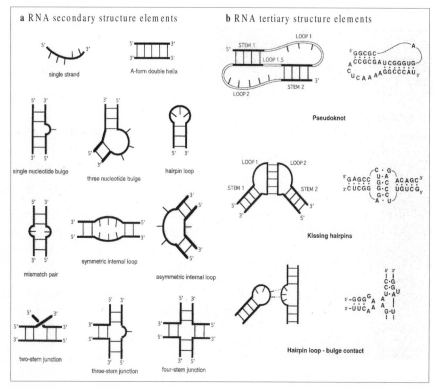

Fig. 2.1a RNA secondary structure elements. Secondary structures can be divided into single-strand regions, helices, bulges, hairpin loops, internal loops and junctions. These motifs are combinations of Watson-Crick base pairs, non-Watson-Crick base pairs (mismatch pairs) and unpaired bases. Watson-Crick base pairing brings two complementary RNA single strands together to form a right-handed double-helix with an A-form geometry (Saenger, 1984). The distinction between a single-strand region and a bulge, loop or junction is that in a single strand the ends are not constrained. In contrast, the ends of bulges, loops or junctions must be in a tightly limited volume. Single-strand regions next to helices are dangling ends; the dangling nucleotide may be on a 5' end or a 3' end. (After: Nowakowski and Tinoco, 1997). **b** RNA tertiary structure elements. Tertiary structures are formed as higher order structures from interacting secondary structure elements. (a) A pseudoknot consists of hairpin loops and single-stranded regions, it is formed when nucleotides from a hairpin loop pair with single-strand elements adjacent to the stem of the hairpin. The specific example shown, the pseudokont causes frameshifting in mouse mammary tumor virus (Shen and Tinoco, 1995)

2.2 The Different Types of RNA and Their Regulatory Elements

We can group the RNAs with regard to their localization, function and abilities in different families, which are summarized in Table 2.1, and described in more detail in this chapter as well as in Chapters 5 and 6. The remaining chapter first explains different regulatory RNA motifs in more detail and example structures are represented in figures.

Table 2.1. Different types of RNA and regulatory motifs

Nuclear RNAs, their metabolism and function

- *Splicing of pre-mRNAs to mRNAs* (Madhani and Guthrie 1994; Lamond 1995; Burge et al. 1999)

The splicing of pre-mRNAs, the products of RNA polymerase II, to mature mRNAs is a two-step process, involving transesterefication reactions, in which first a 2' - 5' bond is generated at the branch point, upstream of the 3' splice site. This results in the formation of a lariat RNA intron intermediate with a free 3' OH at the 5' exon, which is released when the phosphodiester bond at the 3' splice is broken in the second step of the reaction. The splicing reaction and the final joining of the exons occurs within the spliceosome.

- Splicing motifs

5' and 3' splice site consensus, branch point consensus

- *Small nuclear RNAs* (snRNAs)

snRNAs form together with common small proteins small nuclear ribonucleoproteins (snRNPs), which are the core of the spliceosome U1, U2, U4/U5/U6 (Nilsen 1994a; Lamond 1995; Yu et al. 1999). With specific proteins, less abundant species, e.g. U41, U42 in *S.pombe,* are assembled (Dandekar and Tollervey 1989; Yu et al. 1999)

- *Self-splicing introns* (Gilbert and de Souza 1999; Lambowitz et al.; 1999)
  -*Tetrahymena* intron (Cech 1993; Cech and Golden 1999)
  -Group I and Group II introns (Lambowitz et al. 1999)

- *Trans-splicing RNAs* (Nilsen 1995)

Trans-splicing RNA substrates are a small capped, but non-polyadenylated RNA, the spliced leader (SL) RNA containing the 5' exon (SL) and splice donor site, and the acceptor pre-mRNA, which contains the splice acceptor and 3' exon. *Trans*-splicing occurs in nematodes (*Trypanosoma, Ascaris, Caenorhabditis*), trematodes *(Schistosoma)* and protists *(Euglena)* and cestodes *(Echinococcus)* (Nilsen 1994b, 1995; Blumenthal 1995; Brehm et al. 2000). Extensions to other organisms are discussed in Chapter 6.

• *RNA editing* (Simpson 1999; Gott and Emeson 2000)
RNA editing is defined as site-specific insertion, substitution or deletion of nucleotides in RNAs. It has been described for organisms from protozoa to man, first for the mitochondrial genome (kinetoplast) in trypanosomes, in physarium mitochondria, plant mitochondria and chloroplasts (Simpson and Thiemann 1995). Besides kinetoplast RNA, editing can affect mRNA, rRNA and tRNA.

- mRNA editing
Cytochrome oxidase subunit II mRNA in *Trypanosoma* kinetoplasts (Benne et al. 1986) was the first example for which the term RNA editing had been introduced. A cleavage-ligation mechanism involving a guide RNA (gRNA) as template providing the editing information (Blum et al. 1990) could be identified experimentally (Hajduk et al. 1993; Stuart et al. 1997; Hajduk and Sabatini 1998; Estevez and Simpson 1999). mRNA editing results in new start and stop codons (Stuart et al. 1997; Chang et al. 1998; Steinhauser et al. 1999) frameshifting between alternative open reading frames (Kolakofsky and Hausmann 1998), splice-site alterations (Rueter et al. 1999), amino acid substitutions (Nagalla et al. 1994; Herb et al. 1996; Skuse et al. 1996).

- tRNA editing
Editing in tRNA creates essential structural elements at the primary, secondary and tertiary levels, involving both loop nucleotides and base-paired stems (Price and Gray 1998). It can occur by nucleotide deletion/insertion (Price and Gray 1999) nucleotide insertion (Antes et al. 1998), or base conversion (Borner et al. 1996). As a result of the editing the recognition of the tRNA by aminoacyl synthetase can be altered, 5' and 3' end processing or subsequent base modifications can be affected (Borner et al. 1996; Marchfelder et al. 1996; Fey et al. 2000). A large number of bases in tRNA are subjected to editing (Limbach et al. 1994), including the four canonical nucleotides. Examples have been described in *Acanthamoeba* (Lonergan et al 1993), *Spizellomyces* (Laforest et al. 1997), myxomycetes (Antes et al. 1998), land plants (Binder et al. 1994), metazoans (Yokobori and Paabo 1995), squid (Tomita et al. 1996) and other organisms, including bacteria.

- rRNA editing
rRNA editing, although of functional importance, appears to be less frequent than editing in mRNA and tRNA. In *Dictyostelium* mitochondrial small subunit rRNA a single nucleotide exchange has been found in the highly conserved 530 loop,

which is thought to play a critical role in both, tRNA selection and proofreading (Powers and Noller 1994; Barth et al. 1999). The same motif, as well as along the entire small and large rRNAs in *Physarium* mitochondria, contains nucleotide insertions (Miller et al. 1993; Mahendran et al. 1994). These changes, due to editing, can be predicted to affect translational fidelity.

The RNA moiety in telomerase *(Blackburn, 1999)*
Telomerase consists of an RNA and protein moieties required for the elongation of the telomer strand. In *T. termophila*, the first telomerase RNA (TER) of 159 nucleotides was found (Greider and Blackburn 1987, 1989). *Tetrahymena* TER, like other small eukaryotic RNAs, is an RNA polymerase III transcript and bears no 5' cap.
In contrast, human (~450 nts), mouse (~400 nts) and yeast TER are RNA polymerase II products (Blasco et al. 1995; Feng et al. 1995; Chapon et al. 1997). TER is associated with the telomerase reverse transcriptase (TERT) and functions as a matrix for the reverse transcription reaction catalyzed by telomerase. In *Tetrahymena* a RNP complex consisting of TER and TERT could be identified (Bhattacharyya and Blackburn 1997)

- *Ribosomal RNA processing* (Venema and Tollervey 1999)

Primary rRNA (pre-rRNA) is transcribed by RNA polymerase I. Pre-rRNA transcription takes place in the nucleolus. The pre-rRNA genes function as nucleolar organizers by localizing the transcription machinery and the components required for ribosomal subunit formation (Karpen et al. 1988). The pre-rRNA is extensively processed, nascent pre-rRNAs are immediately associated with proteins forming pre-ribonucleoprotein particles (pre-rRNPs). The positions of pre-rRNA cleavage sites and specific modifications, i.e. 2'-O-methylation (Kiss-László et al. 1996) and pseudouridinylation (Ganot et al. 1997; Ni et al. 1997) are determined by more than 150 small nucleolar RNAs (snoRNAs). The snoRNAs hybridize transiently with the pre-rRNA and form, like snRNAs (see above), with associated proteins, snoRNPs (Morrissey and Tollervey 1995).

- *Ribosomal RNA (rRNA)*

The final products of pre-rRNA (45S or 13.7 kb) processing in higher eukaryotes are 28S, 18S and 5.8S rRNA. The 5S rRNA is transcribed by RNA polymerase III in the nucleoplasm, the transcript does not require processing, and assembles, after diffusion to the nucleolus, with the 28S, and 5.8S and rRNAs into the large 60S ribosomal subunit. The 18S rRNA is a component of the small 40S subunit. The rRNA constitutes about two thirds of the mass of ribosomal subunits. In *S. cerevisiae* the primary pre-rRNA (35S or 6.6 kb) is processed to 25S, 18S and 5.8S rRNA, a 5S rRNA is transcribed separately. In *E. coli*, 23S, 16S and 5S mature rRNAs exist; there is no prokaryotic homologue of eukaryotic 5.8S rRNA (Chu et al. 1994; Venema and Tollervey 1995).
Although the rRNA species differ in size between mammals, yeast and bacteria,

functional important structures, e.g. stem loops, are conserved.

- *Small nucleolar RNA (snoRNA)* (Balakin et al. 1996)

snoRNAs are either transcribed by RNA polymerase II and III from their own promoter or they are spliced-out introns of genes encoding functional mRNAs. Remarkably, some snoRNAs are introns spliced out from apparently non-functional mRNAs. It seems that the genes encoding these mRNAs exist only to express snoRNAs from excised introns. A large class of snoRNAs, which is involved in rRNA 2'-O-methylation by snoRNPs, bears sequences by which it binds to fibrillarin, a nucleolus-restricted protein. In addition, it contains a sequence thought to bind a methyltransferase, which catalyzes this modification. RNase MRP, another snoRNP, catalyzes the cleavage of the spacer sequences in the pre-mRNA. The snoRNA associated with RNase MRP is homologue to the RNA of RNase P, which is mainly involved in tRNA processing. There is evidence that not only RNase MRP, but also RNase P catalyzes pre-rRNA cleavage (Chu et al. 1994; Lund and Dahlberg 1998).

Cytoplasmic RNAs and their regulatory motifs

- *Transfer RNAs (tRNAs)*

Cytoplasmic tRNAs
Mitochondrial tRNAs
Chloroplasts tRNAs

- **Posttranscriptional regulatory signals in mRNAs**

Examples of motifs in the coding and non-coding sequences of mRNAs, which are important for the posttranscriptional regulation of gene expression will be described in detail in Table 2.2.

- Further cytoplasmic RNAs

7SL RNA, signal recognition particle (SRP) (Lütcke 1995)
10Sa RNA (also termed transfer-messenger RNA; tmRNA); Jentsch 1996).

Catalytic RNAs

Hammerhead ribozymes, and other ribozymes, the *Tetrahymena* self-splicing introns, RNase P are described as examples for catalytic RNAs in this chapter (Symons 1992; Long and Uhlenbeck 1993; Hertel et al. 1996; Cech 1993; Lohse and Szostak 1996).

Viral RNAs

Table 2.3 gives a comprehensive summary of viruses with RNA genomes, specific examples of regulatory motifs in viral RNA sequences are discussed in this chapter.

General Nuclear RNA-Binding Motifs and Nucleoproteins

In the metabolism of the cell nucleus, different RNA-binding motifs are required for RNA transport and RNA processing to target the interaction between RNA and nucleoprotein. Examples include RNA transport by specific proteins such as the NPL3 protein (Flach et al. 1994). The important question of which RNA motifs are recognized has still to be examined in more detail, for instance by affinity selection schemes (Chap. 3).

Small RNAs are recognized by nucleolin, a major nuclear phosphoprotein, which is probably involved in rDNA transcription, rRNA packaging and ribosomal assembly. Each of these steps requires accurate targeting by suitable and as yet partially uncharacterized RNA motifs (Schmidt-Zachmann and Nigg 1993). Another example is the nuclear localization signal or sequence (NLS) which targets RNAs and proteins to the nucleus. The protein signal consensus is a bipartite signal (Robbins et al. 1991). General principles of RNA recognition by proteins are apparent in consensus RNA recognition motifs (Brand and Bourbon 1993). Thus arginine-rich motifs (e.g. the hepatitis delta antigen; Lee et al. 1993) provide via the arginine fork a general way to recognize different RNAs: an arginine residue may contact by the amines in the fork-like structure of its side chain RNA. The RNA recognition code in the arginine-rich proteins seems to be otherwise surprisingly flexible (Dandekar and Argos 1996; Harada et al. 1996). For many other interactions the amino acid requirements are more complex, for instance in the RNA-protein interactions involved in the regulation of *courtship song* in *Drosophila* (Rendahl et al. 1992). Table 2.4 gives a short summary of different amino acid and protein motifs involved in RNA recognition.

Specific signals encoded in viral RNAs are also mentioned here. Examples are: HIV in which the early regulatory protein Tat binds the *trans*-activation response element (TAR) in the long terminal repeat (LTR) of the viral RNA or the interaction between the rev-response element in the HIV RNA and -protein (Cullen and Malim 1991; Dandekar and Koch 1996).

The polyadenylation signal for 3' end maturation of the mRNA precursor in the nucleus seems at first glance to be a rather simple RNA motif (AAUAAA; Table 2.2). However, different proteins recognize and specifically interact with this regulatory signal and in addition, more discrete motifs are involved in recognizing polyadenylated RNA in different steps (Manley 1995). Poly(A)-binding protein is an essential component of the eukaryotic cell. Apart from those proteins required for the transport of the mature RNA from the nucleus to the cytoplasm there are specific proteins involved in the polyadenylation process itself. The mRNA motif

AAUAAA was the first specific RNA signal revealed, but current work shows several additional RNA signals recognized (Keller 1995; Beyer et al. 1997).

However, completely different proteins also recognize this regulatory mRNA signal, such as *rox8* in *Drosophila,* which binds to poly(A) sequences to trigger apoptosis (Brand and Burbon 1993).

The basic signal of poly-uridylated RNA is also recognized by specific proteins, for instance the La mouse antigen. As discussed above for the poly-adenylation signal, the complex motifs responsible for specifically targeting different RNA-protein interactions are hidden in the poly(U)-stretches and variations of RNAs recognized by the La mouse antigen and various other proteins. Several heteronuclear ribonucleoproteins (hnRNPs) bind specifically to distinct pre-mRNA motifs.

2.3   RNA Motifs Involved in Splicing

Before the mature mRNA appears, its precursor has to be processed. The basic pre-mRNA splicing reaction is outlined in Fig. 2.10, but the details of processing are complex (Crick 1979; Chambon, 1981; Rymond and Rosbash 1992; Higgins and Hame 1994; Madhani and Guthrie 1994; Nagai and Mattaj 1994; Newman 1994). However, for our purpose of identification of regulatory RNA by virtue of its structures and regulatory motifs, for instance to reveal or identify new instances of snRNAs, several different motifs are exemplified here. A compilation of the principal snRNAs is found in Table 2.2. In Figures 2.10-2.13 several typical features are highlighted. Besides the high amount of secondary structure present in each of them, there are specific snRNA-snRNA and pre-mRNA interactions (U4/U6, Fig. 2.11; U6/U2, Fig. 2.11; U4/U5/U6, interactions of U5 with pre-mRNA, Fig. 2.11; interactions of U1/U2/U6, Fig. 2.12). The structure of most snRNAs is highly conserved between distant species. The homology is high enough to allow identification in practice by fast motif-searching procedures such as BLAST or BLAST2 (BLAST2 is an optimized search, allowing for gaps and deletions; see Chap. 4). However, the identification of new and related snRNAs can also be attempted by exploiting conserved motifs such as the Sm site which is the motif AGUUUUGA, more generally purine, purine, poly-uracil, purine, purine (Lamond 1995). This RNA motif in the snRNA is critical for assembly of the proteins common to all spliceosomal snRNAs. These proteins assemble in the cytoplasm in the snRNA region of the Sm site. The anchoring of the attached proteins to the RNA motif is also important for directing the re-import into the nucleus of the snRNAs with attached proteins (Görlich and Mattaj 1996). Selection of alternative sites for splicing (Fig. 2.13) leads to different RNA messengers and proteins and is important in tissue-specific differentiation and development.

Variants of the standard splicing complex have, in fact, been already identified. One is the U1-less splicing found in *Trypanosomes* and related organisms (Nilsen

1995). Important motifs include the stem-loop structure formed by the SL RNA. This structure provides the *trans*-spliced leader and is helpful for the identification of these RNA species in other organisms, whilst in addition shedding light on the splicing reaction of these variants (Chap. 4). A recently characterized variant of splicing utilizes a modified spliceosome involved in splicing of ATAC introns (Nilsen 1996)

For the function of splicing or indeed other RNA processing reactions such as rRNA processing, the RNA structure does not operate alone, but rather the catalytic or processing reactions are enhanced by interaction with proteins. These interactions probably originally evolved in order to process mRNA faster and more precisely than would be possible with RNA alone (Gesteland and Atkins 1993). However, the co-evolution of protein factor and RNA has in many instances continued for so long that the protein has become an essential factor for the reaction (Lamond 1995; Göhrlich and Mattaj 1996). One example of this close interdependence is the spliceosome, in which the common proteins B, B', D1, D2, D3, E, F and G are included in each of the snRNPs formed with one of the central splicing snRNAs U1, U2, U4, U5 and U6 as the RNA core. Subsequently, protein factors may evolve to adopt additional functions as the splicing factor U2AF, recently found in *C. elegans,* which facilitates the interaction of the U2 snRNP with the splicing branch point. Further examples include the protein factors recognizing the Sm site and which are parts of the spliceosome. Today, seven Sm proteins (B/B', D1, D2, D3, E, F and G) are known and their structure has been determined (Kambach, et al. 1999). In addition, they are involved in regulation of snRNA transport. Different snRNA specific proteins developed such as 70k, A and C protein binding to different single-stranded loop regions of snRNA U1 or A' and B" in snRNA U2. Autoregulatory interactions evolved such as U1A protein with its own mRNA (Fig. 2.2). Finally, modifiers of the splicing process and mRNA precursor export exist, such as TAR and rev protein. These interactions also require specific RNA sequence and structure.

These specific motifs can be exploited to suggest novel candidates for regulatory interactions, for instance in the case of the rev-response element a related structure occurs in the messenger RNA of human, mouse and DNA ligase I (see Chap. 6; Dandekar and Koch 1996).

Besides the snRNAs there have been a number of snRNPs characterized as hnRNPs. In addition to processed introns and their associated protein-binding factors, these yet only partially or incompletely characterized RNAs are good candidates for the detection of new regulatory RNA structures.

After successful splicing, the 3' end of the pre-mRNA is cleaved, and polyadenylation takes place. These processes are again directed by specific RNA signals.

An intronic splicing silencer that causes skipping of the IIIb exon of fibroblast growth factor receptor 2 through involvement of polypyrimidine tract binding protein has been identified (Carstens et al. 2000). Several intron elements that contribute to exon 7b skipping, resulting in the production of two proteins from hnRNP A1 pre-mRNA: hnRNP A1 and the less abundant A1B, have recently been described Simard and Chabot (2000).

Self-Splicing Introns

Several RNA sequences do not depend on the complex machinery of the spliceosome for removal of their introns. Their RNA structure enables autocatalytic removal of the intron. Two different classes, group I (Fig. 2.25) and group II (Fig. 2.26) are apparent, the reaction mechanism is generally similar to that shown for spliceosomal splicing in Fig. 2.10.

Trans-Splicing

In this variation of the splicing reaction a small RNA sequence, the leader, is passed (or transferred, hence *trans*-splicing) from one RNA (the *trans*-spliced leader RNA) to another RNA, generally a messenger RNA. Depending on the organism, this process can occur in most mRNAs in the cell (*Trypanosomes*) or only a group of them (Nematodes) (Laird 1989; Nilsen 1995). Several *trans*-spliced sequences are shown in Fig. 2.14. The consensus formulated from these enabled us to retrieve all the known instances of *trans*-spliced leader carrying RNAs and organisms. In addition this analysis revealed the occurence of *trans*-splicing in other organisms (Dandekar and Sibbald 1990). Recent developments confirm *trans*-splicing in higher organisms.

2.4   RNA Editing

A further complication is that several mRNA transcripts undergo the addition of nucleotides or modification of their sequence, though this has not been encoded in their coding genomic DNA sequence. This process is called editing. Several examples are shown in Table 2.2; further details are given in Chapter 6 and Fig. 6.2.

- **Motifs and RNAs Involved in Ribosomal RNA Processing**

Ribosomal processing is a reaction of comparable complexity to splicing; however, speed is even more critical, as such a high quantity of rRNA is synthesized in most living cells. Thus, the system has evolved several proof-checking mechanisms, and redundancy. Due to their abundance, rRNA species for most organisms have been characterized. Detailed structure and motifs are not given here; instead, the reader is referred to the review by Venema and Tollervey (1999) and to Chapter 6.

Structure and motifs are tightly connected to the requirements of ribosomal architecture, rapid processing and synthesis and efficient protein translation. Figure 2.2 illustrates specific RNA structures in ribosomal RNA to which the small subunit ribosomal protein S8, and the large ribosomal subunit protein L11 bind and which are involved in translation. The ribosomal RNA itself is far more active in the ribosome than previously thought (Venema and Tollervey 1995, 1999). The loop motif, shown in Fig. 2.2 illustrates that also in the process of rRNA maturation and processing RNA-protein interactions are directed by specific RNA signals. An example of such an interaction is nucleolin which binds the 5' external transcribed spacer (5' ETS) of the long pre-rRNA transcript. Nucleolin is a principal component involved in establishing the nucleolar architecture where the pre-rRNA is transcribed and processed. As rRNA synthesis occurs in the nucleolus, a dense sub-compartment in the nucleus of the cell, the different RNAs involved in processing site selection are called snoRNAs. Due to the requirement of speed and redundancy in processing of the rRNA precursor in many of the different processing steps, several snoRNAs and proteins interact with rRNA (as do snRNAs and many proteins in the maturation of the pre-mRNA). Specific, sometimes complex and/or not yet characterized rRNA motifs direct cleavages. Some of the cleavage steps and components involved are shown, outlined in Fig. 2.15, and the different classes of snoRNAs and functions are summarized. Consensus structures for the two major classes, ACA and box C/box D allow the identification of new snoRNAs (Chap. 4 for consensus searches and Chap. 3 for experimental tests). Figure 2.16 shows direct base pairing interactions between rRNA and snoRNAs, which serve as methylation guide RNAs for rRNA methylation (Tollervey and Kiss 1997). A consensus structure can be built and applied to identify new snoRNAs of this type (see Chap. 6; Seraphin 1993; Kiss-László et al. 1996), several of which are, in fact, intron-encoded (Chap. 6; Kiss-Laszló et al. 1996). The discovery of new snoRNAs has been facilitated by the availability of the complete sequence of the yeast genome, and is a good case in point for designed database searches for regulatory RNA motifs (Chap. 4). Using this consensus structure over 200 additional homologues of snoRNAs, so-called sno-like RNAs (sRNSs) were identified in *Archaea* (Gaspin et al. 2000).

2.5   Posttranscriptional Regulatory Signals in mRNA

- *mRNA 5' UTR*

Regulatory signals in the 5' UTR of the mRNA are often involved in regulation of translation and its initiation.

A well-characterized example from prokaryotes is translational regulation by the GCN4 element. The GCN4 gene encodes a general activator for amino acid synthesis. Short upstream open reading frames modulate ribosomal initiation at the long open reading frame; thus, these RNA signals effect translational regulation (Chaps. 3 and 6; Müller and Hinnebusch 1986, Hinnebusch 1997).

A variation on the theme of translational regulation is the regulation of mRNA translation by the IRE in the 5' UTR of, for instance ferritin mRNA (Fig. 2.5). The binding of the protein partner, the iron-response element binding-protein (IRP) to the IRE prevents the binding of the 43S translation pre-initiation complex and hence blocks translation initiation (Gray and Hentze 1994; Muckenthaler et al.

1998). Higher levels of iron lead to the formation of the 4Fe/4S iron-sulfur cluster in the IRP, causing the releases of IRP from the IRE in the 5' UTR and abolishing the translational block. This regulation is utilized in several proteins, which are expressed when higher iron concentrations are present in the cell. Thus further iron is stored by ferritin in the cell after ferritin mRNA.

- Open Reading Frame

The structure of the mRNA within the open reading frame exerts several regulatory effects. There is a clear influence on translational speed and, hence, on protein folding. It is possible to compile examples of mRNA where a pause in translation caused by mRNA structure is important to achieve correct protein folding (Brunak and Engelbrecht 1996; Thanaraj and Argos 1996). Further RNA signals, for instance those conferring stability, can also be present in this region and thus several codes overlap with that encoding the amino acid sequence (Chap. 4).

- *mRNA 3' UTR*

RNA stability and pre-mRNA processing are often regulated by motifs and signals present in the mRNA 3' UTR. The 3' UTR serves as a switch for directed release of information; thus a stability signal in the 3' UTR may determine that a certain protein can be translated only for a limited time.

In addition, pattern formation in the early development of many organisms relies on localized cytoplasmic proteins, which can be pre-localized as mRNAs. The *Drosophila* gene *oskar*, required both for posterior body patterning and germ cell differentiation, encodes one such mRNA. Localization of *oskar* mRNA is an elaborate process (Erdelyi et al. 1995) involving movement of the transcript first into the oocyte from adjacent interconnected nurse cells and then across the length of oocyte to its posterior pole. Several elements within the *oskar* 3' UTR affect different steps in the process: the early movement into the oocyte, accumulation at the anterior margin of the oocyte and finally localization to the posterior pole; again, this may be a more general principle (Kim et al. 1993). Another example in *Drosophila* embryos is the graded activity of the posterior determinant *nanos* (*nos*). It generates abdominal segmentation by blocking protein expression from maternal transcripts of the *hunchback* (*hb*) gene. When expressed ectopically at the anterior pole, *nos* can also block expression of the anterior determinant *bicoid* (*bcd*). Both regulatory interactions are mediated by similar sequences in the 3' UTR of each transcript (Wharton and Struhl 1991). These *nos* response elements (NREs) are both necessary and sufficient to confer *nos*-dependent regulation, the degree of regulation being determined by the number and quality of the elements and the level of *nos in vivo*. Nos acts as a morphogen, controlling *hunchback* expression (and hence the abdominal pattern) as a function of its concentration-dependent interaction with the NREs (Fig. 2.6).

There are at least three well-known types of functional signals in the 3' UTR: Besides mRNA localization signals (Fig. 2.6), best exemplified by developmental genes, there are various polyadenylation signals (Figs. 2.2 and 2.3). Finally the regulation of mRNA stability is often mediated by signals encoded in this region. Figure 2.7 shows the 3' UTR region of the transferrin receptor mRNA as an example (explained below).

Stability and decay of mRNAs is regulated by different signals and motifs in the RNA structure. RNA signals during development are reviewed in Surdej et al. (1994), such as the CU-rich stability element in the 3' UTR of alpha-globin mRNA (Kiledjian et al. 1995). However, there are also different metabolic pathways controlled by the regulation of translation using RNA motifs in the 3' UTR to which regulatory proteins bind. An excellent example for translational control from the 3' UTR is the reticulocyte-15-lipoxygenase (LOX) mRNA (Ostareck-Lederer et al. 1994; Ostareck et al., 1997, 2001; Fig.2.3). The LOX protein is expressed in erythroid cells only just before they become mature erythrocytes and mediates mitochondrial breakdown. This temporal restriction is achieved by translational silencing of the LOX mRNA in erythroid precursor cells. The LOX mRNA 3' UTR differentiation control element (DICE) binds the KH-domain proteins, hnRNP K and hnRNP E1, which leads to the establishment of a translationally silenced mRNP. An important part of the mechanism of translational silencing has very recently been resolved. The hnRNPs K and E1/DICE complex formed at the 3' UTR inhibits the joining of the 60S ribosomal subunit to the 43S pre-initiation complex, which has reached the initiator AUG (Ostareck et al. 2001).

The stability of the transferrin receptor mRNA is regulated by RNA secondary structure motifs (Fig. 2.7) in its 3' UTR (Müllner and Kühn 1988). This exemplifies again the context specificity of regulatory RNA signals (Chap. 1): in the 3' UTR of an mRNA context, multiple copies of the IRE, if used as a regulatory RNA structure, are not interpreted as a translational signal but as a stability modulator instead. The binding of the IRP to the IRE is dependent on the cellular iron content. Under iron starvation, IRP binds to the messenger RNA and its elements in the 3' UTR, thus stabilizing it, with the effect that more transferrin receptor is synthesized, to import more iron into the cell. Under conditions in which the cellular iron level is high, the formation of a 4F/4S iron-sulfur cluster in the IRP causes a conformational change in the protein such that it is no longer able to bind the IRE. As a consequence, the messenger is more rapidly degraded, less transferrin receptor is synthesized and the iron level is downregulated. The IRE signal is interpreted differently in the context of the 5' UTR (see next section).

mRNAs encoding selenocystein (Se-Cys) containing proteins bear specific Se-Cys insertion elements (SECIS) in their 3' UTR (Fig. 2.4). These SECIS elements allow the incorporation of Se-Cys by recoding an UGA stop codon. The mechanism by which this is achieved is discussed in detail in Chapter 6.

The cytoplasmic polyadenylation element (CPE), a *cis* element in the 3' UTR of cyclin B1 mRNA, mediates translational repression in GV-stage mouse oocytes (Tay et al. 2000). Histone mRNA 3' end formation is an essential regulatory step producing an mRNA with a hairpin structure at the 3' end. This requires the interaction of the U7 snRNP with a purine-rich spacer element and of the hairpin-binding protein with the hairpin element, respectively, in the 3'UTR of histone RNA (Muller et al. 2000).

2.6 Further Regulatory RNAs

Interestingly, translation initiation and elongation may also be regulated not by the mRNA structure itself but by other regulatory pathways involving regulatory RNA. Calcium depletion activates the double-stranded RNA-dependent protein kinase (PKR) to phosphorylate the alpha subunit of eIF-2. This activation is likely to be mediated through the PKR RNA-binding domain, since overexpression of the domain is sufficient to inhibit increased phosphorylation of the alpha subunit of eIF-2 upon A23187 (an ionophore for calcium depletion) treatment. Therefore PKR is regulated by double-stranded RNA (Srivastava et al. 1995; Kaufman 2000). There are parallels in prokaryotic systems. The translational repressor S4 in *E.coli* binds to the alpha mRNA (Fig. 2.8). A very stable 8-bp helix forms upstream from the ribosome-binding site in this mRNA, defining a 29 bp loop which is altered upon protein binding to mediate translational repression (Deckman and Draper 1987). Further examples include the attenuation of translation in the His operon and Trp operon in *E.coli* (Yanofski et al. 1996).

The gene-5 protein of filamentous bacteriophage fd is a single-stranded DNA-binding protein. It binds non-specifically to all single-stranded nucleic acid sequences and is capable of binding specifically to the sequence d(GT(5)G(4)CT(4)C) and their RNA eqivalent r(GU(5)G(4)CU(4)C). Translational repression occurs when an approximately 170 kDa nucleoprotein complex consisting of four oligonucleotide strands and eight gene-5 protein dimers is formed (Oliver et al. 2000).

2.7 Modified Residues in Eukaryotic mRNAs

The formation of internal 6-methyladenine residues in eukaryotic mRNA is a post-translational modification in which S-adenosyl methionine serves as a methyl donor; 30-50% of methyl groups incorporated into mRNA bear m_6A residues. Although most of the cellular mRNAs and certain viral mRNAs contain at least one m_6A, some transcripts, such as those coding for histone and globin, are devoid of such modifications. m_6A modifications have also been found in heteronuclear RNA and are conserved during mRNA processing. In all cases a strict consensus sequence is found with Gm_6AC or Am_6AC in the transcript.

Although the biological significance of the modification remains unclear, it is suggested that modification may be required for mRNA transport to the cytoplasm, the selection of splice sites or other RNA processing reactions (Tuck 1992).

2.8 Correction of Missing Stop Codons by 10Sa-RNA

10Sa RNA has the interesting function of labelling the protein product derived from an mRNA which has a missing stop codon (Fig. 2.9). Such mRNAs, which are partly deleted by nucleases from the 3' end, give rise to incomplete proteins and are swiftly degraded by the cell. The ribosome is normally stalled on such a messenger. The arrival of 10Sa RNA releases such a stalled ribosome due to its tRNA-like shape (Fig. 2.9). By a translational switch, an amino acid tag is subsequently translated from the 10Sa RNA and attached to the partly synthesized protein. Thus 10Sa RNA, also termed transfer-messenger RNA (tmRNA). The tag on the protein is rapidly recognized by periplasmic tail-specific proteases and thus the incompletely synthesized protein is efficiently degraded (Jentsch 1996). This is an exciting example of a novel regulatory RNA involved in correcting premature termination in mRNA; however, there are in addition further well-known mechanisms to counteract premature stop codons such as nonsense suppression (Kuchino and Muramatsu 1996) and degradation of mRNA (Atkin et al. 1997).

2.9 Direct Regulation of Gene Expression

Autoregulatory RNA

Examples are accumulating in which the protein encoded by an mRNA autoregulates its synthesis, e.g. by influencing the stability of its own mRNA (Kyrpides and Ouzounis, 1993). These authors suggest that strict autoregulation is an atavism indicating a first mode of the protein-RNA interaction in the RNA world. Whilst we consider the model to be speculative (the following examples may be fine-tuned adaptations acquired recently or a mixture between both extremes), these RNAs are interesting in their own right and include examples from different mRNA regions:
- An RNA transcription factor, composed exclusively of RNA (Young et al. 1991).
- Splicing of its own mRNA:
In a yeast maturase (Lazowska et al. 1980), yeast L32 ribosomal protein (Dabeva et al. 1986), *Xenopus laevis* L1 ribosomal protein (Caffarelli et al. 1987) and *Drosophila* supressor of white *apricot* gene (Zachar et al. 1987).
- Interaction with a 5' UTR:
Preproinsulin interacts with its own mRNA (Knight and Docherty 1991); polysomal protein with chicken vitellogenin mRNA 5' UTR (Liang and Jost 1991)
- Regulation of polyadenylation of the own mRNA:
U1A protein inhibits polyadenylation of its own mRNA to allow a stable level of its own expression (Boelens et al. 1993; Fig. 2.2)
-Regulation of the stability of its own mRNA:
L2 ribosomal protein in yeast (Prescutti et al. 1991); *Bacillus subtilis* ribosomal

protein S4 of the *rps* gene (Grundy and Henkin 1991), *per* gene of *Drosophila* (Hardin et al. 1990), H4 histone gene (Peltz and Ross 1987) and beta-tubulin gene (Gay et al. 1989) and other examples for stability of other mRNAs than its own.
-Translation of its own mRNA:
E.coli ribosomal proteins (Draper 1987), phage T4 gene 32 protein (von Hippel et al. 1982), human thymidylate synthase mRNA (Chu et al. 1991)
-Posttranscriptional regulation of yeast MET2 (Forlan et al. 1991).

Antisense RNA

Antisense RNA (aRNA), which is encoded in *trans*, can regulate gene expression (Delihas 1995; Fig 2.18). Often aRNA is only partially complementary to its target RNA. A YUNR motif is ubiquitous in such system. It forces the building of a sharp bend. The result is a U-turn structure, which provides a scaffold for rapid interaction with complementary RNA (Franch et al. 1999). For example, *micF* in *E.coli* and other bacteria controls outer membrane protein F levels in response to enviromental stimuli. *dic F* , also in *E.coli*, is involved in cell division. *Lin-4* is an exciting eukaryotic example of this type of RNA regulation. It is found in *C. elegans* and functions during larval development (Chap. 6). *Sar* phage RNA (Wu et al. 1987; see below) is shown in Fig. 2.18 in the more general context of inhibiting protein expression by aRNA.

The efficiency of conjugation of F-like plasmids is regulated by the FinOP fertility inhibition system. The transfer (*tra*) operon is under the direct control of the TraJ transcriptional activator protein, which, in turn, is negatively regulated by *FinP*, an aRNA, and FinO, a 22 kDa protein.

2.10 Regulation of Prokaryotic Operons

The *Bacillus subtilis* the tryptopan (trp) operon (trpEDCFBA) is regulated by transcription attenuation. Transcription is controlled by two alternative RNA secondary structures, which form in the leader transcript. Thus, this is also an example of specific regulation in the 5' UTR of a prokaryotic mRNA. In the presence of L-trp, a transcription terminator forms, and the operon is not expressed, whereas in the absence of trp, an anti-terminator structure forms, allowing transcription of the operon. The mechanism of selection between these alternative structures involves a *trans*-acting RNA-binding regulatory protein, the product of the mtrB gene, namely the trp attenuation protein (TRAP). The TRAP protein binds to the leader RNA to induce the transcription termination structure in the presence of trp. However, if trp is missing, TRAP does not bind to the mRNA and the anti-terminator is formed (Yanofsky et al. 1996).

A complex polycistronic regulatory locus (*agr),* controls globally the expression of *Staphylococcus aureus* toxins and other exoproteins. *Agr* contains two divergent promoters. One of these promoters directs the synthesis of a 514 nts transcript, *RNAIII*, which acts primarily on transcription initiation (indirectly by means of intermediary protein factors), and, secondarily, in some cases at the level of translation (directly by interacting with target gene transcripts) (Novick et al. 1993; Fig. 2.20).

The copy number of pBR322 is regulated by the regulatory *RNA I*; the origin of replication is complementary to the small aRNA. Interestingly, a G -> T transversion located near its 3' end results in a stable high copy number of approx. 1000 per cell (Boros et al. 1984).

Translational control of insertion sequence 10 from Tn10 occurs by direct pairing between the transposase mRNA and a small complementary regulatory RNA (Simons and Kleckner 1983).

Regulation of gene-expression by RNA structures is not restricted to prokaryotes. An illustrative example of a regulatory RNA for gene expression in plants is the rhizobial RNA from the *sra* gene, which is 213 nts long and forms a stem-loop secondary structure (Ebeling et al. 1991).

2.11 Cytoplasmic RNA Motifs

An important example of a cytoplasmic RNA is the 7SL RNA (Fig. 2.27) as the central part of the signal recognition particle (SRP). Besides 7SL RNA the SRP consists of four proteins (two monomers of 19 and 54 kDa, respectively) and a heterodimer composed of a 68 and 72 kDa polypeptide. 7SL RNA comprises Alu-like interspersed repetitive elements, which are also represented in other classes of RNA such as mRNA, small RNAs and heteronuclear RNA. Evolutionary evidence suggests that the highly conserved 7SL RNA is the parent RNA for these Alu-like elements. In higher eukaryotes, most secretory and membrane proteins are synthesized by ribosomes which are attached to the membrane of the rough endoplasmic recticulum (rER). This allows co-translational movement of nascent proteins across the ER. The ribosomes are directed to the rER by the SRP (Lütcke 1995). The SRP complex binds to the leader signal sequence of the growing peptide chain. The signal sequence-binding subunit of the SRP from *T. aquaticus* contains a hydrophobic compartment that is likely to contribute to the structural plasticity necessary for the SRP to bind signal sequences of different lengths and amino acid sequence. It also consists of an arginine-rich helix-turn-helix motif, which is thought to form the core of the SRP RNA-binding site (Keenan et al. 1998).

Further cytoplasmic RNAs with less characterized function include other RNAs in this size range of 7S (e.g. 7K RNA), and other cytoplasmic RNAs such as Y-RNA (van Horn et al. 1995, thought to be involved in 5S rRNA synthesis).

2.12 Ribozymes and Their Motifs

Catalytic RNA came as a surprise when first discovered in the *Tetrahymena*. There is an increasing number of further examples of artificial and natural catalytically active RNAs (Long and Uhlenbeck 1993; McKay and Wedekind 1999). Examples are viral and other natural hammerhead structures (Fig. 2.21; Symon, 1992), catalytic hairpins (Fig. 2.22) or the RNA component of RNase P involved in tRNA processing (Figs. 2.23, 2.24). Furthermore, there are differently engineered hammer head and hairpin motifs intended for medical applications such as inhibition of unwanted mRNAs (see Chap. 5).

2.13 Viral RNA

In many viruses a compact genome is a selective advantage for survival. Regulatory RNAs and regulatory RNA elements are often found overlapping with protein-encoding sequences to achieve this. The life cycle of the virus is very often intimately linked to the regulatory effect of different regulatory RNA motifs.
-     The interaction of T4 messenger RNAs with the reg A protein is an example of a complex RNA regulatory interaction. The reg A crystal structure is known (Kang et al. 1995). The reg A protein core and beta sheets have similarity to the RNP-1 and RNP-2 RNA binding consensus motif. The RNA-protein interaction represses 35 early T4 phage mRNAs. However, it does not affect nearly 200 other mRNAs. A fascinating unresolved question is thus how the as yet unknown RNA motifs target the interaction to one subset of T4 mRNAs.
-     Several regulatory RNA secondary structure features are present in coliphages (Olsthoorn et al. 1995). In PP7 phage there is a coat protein-binding helix at the start of the replicase gene. Further regulatory structures at the 5' and 3' termini of the RNA are a replicase binding site and the structure of the coat protein cistron start. Some of these features resemble MS2-type coliphages, others the Q-beta type phages. The start of the coat protein gene of RNA phage MS2 adopts a well-defined hairpin structure of 12 bp (including one mismatch) in which the start codon occupies the loop position. The stability of the hairpin is tuned in the sense that it has the highest stability still compatible with maximal ribosome loading (Oolsthorn, et al. 1994).
-     Perhaps the best known viral example is HIV. The gene expression of human immunodeficiency virus type 1 (HIV-1) is controlled quantitatively and qualitatively in large part by the action of two small nuclear viral regulatory proteins termed Tat and Rev (Figs. 1.2, 2.17; Cullen 1994; Cullen and Malim 1991). Tat is a transcriptional *trans*-activator that acts via a structured RNA target sequence (termed *trans*-activation response element, (TAR) to induce high levels of transcription from the HIV-1 long terminal repeat (LTR) promoter element. In addition, the TAR-RNA element interacts with nucleoporins (Fritz et al. 1995). In

contrast, the Rev protein induces the nuclear export of a specific class of viral RNA species that are otherwise sequestered in the nucleus by the action of cellular factors. Like Tat, Rev also interacts with a highly specific *cis*-acting targeting sequence, the Rev response element. Recently, the major packaging signal of HIV-1 has been localized 3' to the major splice donor within the leader sequence. Binding of Gag protein leads to a conformational change, which alters and probably unwinds the RNA structure (Zeffman et al. 2000).

- At this point, it should be emphasized that regulatory RNA signals in hepatitis B viral RNA and hepatitis C viral RNA are at least equally important medically, with more than 1 million deaths per year as a direct or indirect consequence of hepatitis B and C viral infection.

- The shift from viral regulatory to structural gene expression in human T-cell leukemia virus types I and II (HTLV I, II) is mediated by the Rex protein (Black et al. 1994). Rex protein binds an RNA element in the region R/U5 of the 5' long terminal repeat, the Rex-responsive element (RxRE), which contains a specific Rex core binding element (RBE). In this way the virus overcomes inhibition of expression imposed by a contiguous LTR regulatory element (LTR NA). This RNA element contains *cis*-acting repressive sequences (CRS) which are not bound by Rex protein but are bound by a 60 kDa protein (p60 CRS). In addition, there is the regulatory interaction between the U5 RNA and a 40 kDa protein (p40 CRS). Rex binding is regulated by phosphorylation, which could also be a general regulation mode for RNA binding proteins (Green et al. 1992).

- In cells infected by adenovirus type 2 (Ad2), activation of the protein kinase DAI (an interferon-induced defense mechanism) is prevented by the synthesis of a small, highly ordered virus-associated (VA) RNA (VA RNAI) (Ma and Matthew 1993; Fig. 2.17). Comparison of different adenovirus species shows that VA RNA possesses a terminal stem, an apical stem loop and a central domain, although there is large variation in details and size. Conserved tetranucleotides (CCGG:C/UCGG) provide evidence for the apical stem.

- A very interesting feature of several viral genomes is secondary and tertiary structures which allow the cap-independent translation of viral RNA, initiated at internal ribosome entry sites (IRES) using the host cell translation machinery (see Chap. 6). IRES-mediated translation initiation has been extensively studied in several species of the genus Picornaviridae (Pelletier et al. 1988; Pestova et al. 1994, 1996; Belsham and Jackson 2000; Lomakin et al. 2000). Most picornaviruses shut off the host-cell translation machinery by proteolytic modification of key components required for cap-dependent translation of cellular mRNAs, but not by destruction of the host-cell mRNA. Besides other interesting examples, i.e. hepatitis C virus (HCV) IRES (HCV_IRES) (Pestova et al. 1998), the IRES in the intercistronic region (IGR) of the cricket paralysis virus (CrPV) is the currently most striking one, with regard to the limited requirement of components. It has been shown experimentally that this CrPV-IGR-IRES can recruit ribosomal subunits directly, without any initiation factor, to assemble a functional ribosome. Furthermore, an alanine is coded as the first amino acid in the nascent peptide chain (Domier et al. 2000; Wilson et al. 2000a, b). In Fig. 2.19 two IRES structures are shown, the classical swine fever virus (CSFV) IRES (CSFV-IRES) as a representative of the pestiviruses and the CrPV-IGR-IRES.

P22 phage gene expression is also controlled by a small antisense RNA (*sar RNA*), which suppresses the synthesis of the antirepressor protein (active in early infection) synthesis in late infection by being complementary to the ribosome binding site of the mRNA encoding the antirepressor protein *ant* (Wu et al. 1987; Fig. 2.18).

-     Potato virus X requires two stem loop structures (stem loop1 SL1 and stem loop 2 SL2) for expression of plus-strand RNA and its accumulation. Mutational analysis of stem C in SL1 suggests that pairing is more important than the sequence. Nevertheless, in the region of the terminal GAAA tetraloop (a motif known to enhance RNA stability) certain mutations can be detrimental according to the level of RNA accumulation, indicating that this element is important beyond providing structure stabilization (Miller et al. 1998)

-     In rous sarcoma virus (RSV) a ribosomal –1 frameshift event is required for the expression of the Gag-Pol poly-protein at the overlap of the gag and pol open reading frames. The signal for the frameshift consists of two elements, a slippery sequence (AAAUUUA) and a stimulatory RNA structure located immediately downstream. This structure is a complex stem loop with a long stable stem and two additional stem loops. It can form an RNA pseudoknot transiently and the frameshifting caused by the RSV pseudoknot is less dramatic than by other well-known frameshift-inducing RNA pseudoknots (Marczinke et al. 1998).

Liphardt et al. (1999) further describe two different groups of pseudoknots evolved in frameshifting. The first group was found in infectious bronchitis virus (IBV), and possesses a long stem 1 of 11-12 bp and long loop (30-164 nts). The other group is found in mouse mammary tumor virus (MMTV) that contains stems of only 5 to 7 bp and shorter loops. In this MMTV the pseudoknot was kinked at the stem1-stem2 junction, and this kinked formation is essential for frameshifting.

Bacteriophage Q-beta RNA contains an M-site needed for efficient initiation of minus strand synthesis of Q-beta replicase. This M-site consists of two stem-loop structures followed by a bulge loop of unpaired purines. In a mutational analysis severe effects were obtained when one of the helical stems or the unpaired bulge was deleted or substantially shortend. This leads to the conclusion that protein S1 (the alpha subunit of the replicase) functions in the replication of Q-beta RNA by mediating the effect of M-site interaction (Schuppli et al. 1998).

2.14 Other RNA Species

This review has covered only major and well-characterized areas. As hinted for the cytoplasmic RNAs or nuclear RNA species, there are several areas where the RNA species and hence also their regulatory RNA structures are still insufficiently characterized and not yet understood. This situation is not improved by the fact that genome projects often concentrate on the DNA. Northern analysis indicates that many more small RNAs, let alone more rarely expressed RNAs and/ or larger RNAs, are still to be discovered. The Xist RNA (see Chap. 6) illustrates an example of a recently discovered very large RNA with powerful regulatory function.

2.15 A Catalogue of Regulatory RNA Motifs

In Table 2.1 an overview was given indicating major areas where RNA and RNA interactions mediated by specific RNA elements and structures have proven to be important. Table 2.2 and following tables provide a catalogue of many RNAs and RNA elements involved in regulatory function.

Because RNA viruses contain in their small genome a high number of different RNA structures and elements for regulation of their life cycle, a catalogue of RNA viruses follows in Table 2.3.

Several protein motifs which recognize RNA are summarized in Table 2.4, since they provide an indication (at the sequence level) or a starting point (for instance in affinity selection or genetic screens, see Chap. 3) to identify the interacting RNA partner to this proteins.

Table 2.5 summarizes RNA modifications. These are at the same time intermediates in RNA metabolism providing variants of the canonical nucleotides and serve as micro-motifs for defined functions (see Chap. 5 for details). Furthermore, they are easily missed in database searches if one is not aware of nucleotide modifications (Chap. 4).

Table 2.2. Examples of regulatory RNA motifs

•   **RNA and protein interaction motifs of RNAs involved in splicing**
Small nuclear RNAs (snRNAs) -U1, U2, U4, U5 and U6 snRNAs interact by base pairing with each other andspliceosomal proteins (Green 1991; Dreyfuss et al. 1993; Madhani and Guthrie 1994; Newman 1994; Figs. 2.2, 2.10; 2.11-2.13) -U1 snRNA and U2 snRNA base paire with U6 snRNA (Rymond and Rosbash 1992; Fig. 2.12) -U2 snRNA interacts with U6 snRNA (Hodges and Bernstein 1994; Madhani and Guthrie 1994; Figs. 2.11, 2.13) -U4 snRNA base pairs with U6 snRNA (Brow and Guthrie 1988; Madhani and Guthrie 1994; Fig. 2.11) -U5 snRNA loop which mediates base pairing to exon sequences (Madhani and Guthrie 1994). *Interaction of snRNAs with small nuclear ribonucleoproteins (snRNPs)* -U1 snRNA hairpin I binds to U1 70 kDa snRNP (Draper 1995) -U1 snRNA hairpin II binds to U1A snRNP protein (Scherly et al. 1990; Fig. 2.2)

and U2 snRNA hairpin IV binds to snRNP B" (Scherly et al. 1990; Fig. 2.2)

Sequence motifs in RNA substrates
-Pre-mRNA splicing consensus sequences in yeast: 5' AG, branch-point UACUAAC, 3' YAG in yeast (Rymond and Rosbash 1992). In mammals less strict (Madhani and Guthrie 1994).
-Self-splicing introns are found mainly in mitochondria, their autocatalysis requires a complex secondary structure. In Fig. 2.10 the self-splicing pathway is compared to pre-mRNA splicing in Fig. 2.10 (Cech 1993).
-group I introns: 3' G-OH is an essential feature (Cech 1993; Fig. 2.25)
-group II introns: branch site is an essential feature (Michel and Ferat 1995; Fig. 2.26)

trans-splicing reactions *trans*-splice a leader sequence to maturating mRNA precursors. Motif examples are shown in Fig. 2.14, a consensus is shown in Fig. 4.1 (Dandekar and Sibbald 1990; Bohnen 1993; Nilsen 1995).

- **RNA and protein recognition motifs in rRNAs and rRNA processing**

Ribosomal RNA (rRNAs)
-rRNA–S8 (Draper 1995; Fig. 2.2)
-rRNA–L11 (Draper 1995; Fig. 2.2)
-rRNA (5' ETS)–nucleolin (Ghisolfi-Nieto et al. 1996; Fig. 2.2)
Small nucleolar RNAs (snoRNAs)
-snoRNAs are involved in processing and cleavage of rRNA (Maxwell and Fournier 1995; Fig. 2.15). Typical structures and classes are shown in Fig. 2.15. Further functions: intron-encoded snoRNAs are guide RNAs involved in rRNA editing, 2-O-ribose methylation and pseudouridylation (Kiss-Laszlo et al. 1996; Nicoloso et al. 1996; Ganot et al. 1997; Fig. 2.16).

- **RNA motifs involved in regulation of transcription**

-*Bacillus subtilis* tryptophan operon is regulated by two alternative regulatory RNA secondary structures forming in the leader transcript depending on presence or absence of tryptophan (Yanofsky et al. 1996).

-*Staphylococcus aureus* toxins are controlled by a 514 nt transcript, RNA III; transcription regulation by protein factors (Novick et al. 1993; Fig. 2.20)

Motifs in mRNA 5' UTR
-Preproinsulin mRNA 5' UTR mRNA interacts autoregulatorily with own protein encoded (Knight and Docherty 1991).
-Chicken vitellogenin mRNA 5' UTR interacts with polysomal protein (Liang and Jost 1991)

-Human thymidylate synthase mRNA 5' UTR (Chu et al. 1991)
-5' terminal oligopyrimidine tract (TOP) mRNAs 5' UTR coding for numerous ribosomal proteins, i.e. S16, L32 and eukaryotic elongation factors eEF1 and eEF2 (Meyuhas et al. 1996; Avni et al. 1997; Meyuhas and Hornstein 2000).
-regulatory stem-loop binding sites in *Xenopus laevis* ribosomal protein mRNA for nucleic acid binding protein (Pellizzoni et al. 1998).
-Iron-responsive elements (IREs) in the 5' UTR of several mRNAs bind the iron regulatory protein (IRP) (Hentze et al. 1987; Melefors and Hentze 1993; Hentze and Kuehn 1996; Mascotti et al. 1996). The IRE-IRP complex blocks the recruitment of the 40S ribosomal subunit Muckenthaler et al. 1998). Further examples are IREs in the 5' UTR of human and murine ferritin H-chain mRNA, murine and human erythroid aminolevulinic acic synthase (eALAS) mRNA (Cox et al. 1991; Dandekar et al. 1991) porcine heart mitochondrial aconitase mRNA (Dandekar et al. 1991; Gray et al. 1996; Fig. 2.5) and *Drosophila* succinate dehydrogenase mRNA (Gray et al. 1996; Fig. 2.5).
-A 29-base pair loop in alpha-mRNA 5' leader of *E. coli* forms a stable eight-base pair helix to which the repressor S4 encoded by alpha-mRNA binds (Deckman and Draper 1987; Fig. 2.8).

-Upstream open reading frames (uORFs) repress translation in different eukaryotic mRNAs (Morris and Geballe 2000). Examples are:
GCN4 mRNA in *S. cerevisiae* (Müller and Hinnebusch 1986; Hinnebusch 1997). S-Adenosylmethionine decarboxylase mRNA (Mize et al. 1998; Raney et al. 2000) and the oncoprotein MDM2 mRNA (Landers et al. 1997; Brown et al. 1999) in mammals. C/EBPs alpha and beta mRNAs (Lin et al. 1993; Ossipow et al. 1993) in vertebrates.

-*Cellular internal ribosome entry site (IRES) elements* promote translation initiation in the absence of a functional cap-binding protein complex eIF4F (Carter et al. 2000; Holcik et al. 2000). This mechanism allows translation of immunoglobulin-binding protein (BiP) (Macejak and Sarnow 1991) translation in poliovirus-infected cells when overall translation of host cell mRNA is inhibited as a result of eIF4G degradation. Further examples are: fibroblast growth factor 2 (FGF2) (Vagner et al. 1995). Insulin-like growth factor 2 (IGF2) (Teerink et al. 1995). Eukaryotic initiation factor 4G (eIF4G) (Gan and Rhoads 1996; Gan et al. 1998). Ornithine decarboxylase (ODC) (Pyronnet et al. 2000). *D. melanogaster Antennapedia* and *Ultrabithorax* also harbour cellular IRES elements (Oh et al. 1992; Ye et al. 1997).

• Motifs in open reading frames (ORFs)

-Shine-Delgarno sequence in prokaryotes (5-9 nts 5' of the AUG; complementary and bridging to the 3' end of 16S rRNA)

-Premature termination codons (PTCs) in the ORF are recognized by an mRNA decay pathway termed mRNA surveillance (Peltz et al. 1993) or non-sense-mediated decay (NMD). NMD and the cellular factors involved has been extensively studied in *S. cerevisiae* (Culbertson 1999; Czaplinski, et al. 1999; Jacobson and Peltz 2000). In other eukaryotic organisms including mammals, NMD also contributes, directly and indirectly, to the degradation of transcripts bearing premature termination codons (PTCs) (Li and Wilkinson 1998; Hentze and Kulozik 1999; Maquat 2000).

-Protein folding determinants in mRNA sequences can be involved in the formation of alpha- helices and beta-sheets (Brunak and Engelbrecht 1996; Thanaraj and Argos 1996).

-Translational frameshifting, ribosome hopping and read through of termination codons by mRNA secondary structure elements (hairpins and pseudoknots) (Farabaugh et al. 2000)

-In *E. coli*, mRNAs lacking termination codons stall ribosomes. Such missing stop codons are corrected by 10Sa RNA, which mimics a tRNA and allows the release of the stalled ribosome (Komine et al. 1994; Jentsch 1996; Fig. 2.9).

-Rare codons (AAGACAUCUAUAAAAUCUUC) in the *S. cerevisiae* MATa1 transcript promote instability and decay (Caponigro et al. 1993).

- **Motifs in the mRNA 3' UTR**

-The reticulocyte 15-lipoxygenase (LOX) mRNA 3' UTR contains a differentiation control element (DICE), consisting of a ten times-repeated pyrimidine-rich motif (Fig. 2.3). LOX mRNA translation is silenced by hnRNPs K and E1, which bind specifically to the DICE and block translation initiation at the step of 60S ribosomal subunit joining (Ostareck-Lederer et al. 1994; Ostareck et al. 1997, 2001)

-Lactate dehydrogenase A (LDH-A) mRNA 3' UTR contains a protein kinase C (PKC) -stabilizing region (20 nts), which is required for stabilization of LDH-A mRNA by PKC activation in rat glioma cells (Short et al. 2000).

-Translational activation of a selected group of maternal mRNAs, including *mos, cyclin A1* and *-B1*, and *cdk2*. It requires two *cis*-acting 3' UTR elements and three core polyadenylation factors: the *cis* elements are the hexanucleotide AAUAAA (Keller 1995), and an upstream U-rich sequence called cytoplasmic polyadenylation element (CPE) (Fox et al. 1989; McGrew et al. 1989; Richter 1999, 2000). The core factors include a CPE binding protein (CPEB), which contains an RNA recognition motif and zinc fingers (Hake and Richter 1994; Hake et al. 1998), and a cytoplasmic cleavage and polyadenylation specificity factor (CPSF), which interacts with AAUAAA (Dickson et al. 1999). The third core factor is poly(A) polymerase (PAP) (Gebauer and Richter 1995). Mendez et al. (2000a) define the molecular function of CPEB in cytoplasmic polyadenylation (Chap. 6).

-Amastin mRNA abundance (its up-regulation) in the intracellular mammalian amastigote form of *T. cruzi* is controlled by an element in the mRNA 3' UTR. The function of this element, to which a protein complex binds, is position- and orientation-dependent (Coughlin et al. 2000).

-HIV-1 pre-mRNA processing requires a 76 nts element upstream of the 3' UTR core poly(A) site (AAUAAA). This element interacts directly with the 160 kDa subunit of the cleavage polyadenylation specificity factor (CPSF) responsible for the recognition of the AAUAAA hexamer (Gilmartin et al. 1995; Fig. 2.3).

-In the alpha-globin mRNA 3' UTR a pyrimidine-rich element has been identified, which confers mRNA stabilization by binding a protein complex which contains hnRNP E1 (Kiledjian et al. 1995; Ostareck-Lederer et al. 1998). Similar motifs are found in the 3' UTR of tyrosin hydroxylase and alpha-collagen type I mRNA (Holcik and Liebhaber 1997).

-Selenocysteine insertion elements (SECIS) required for the recoding of the UGA stop-codon as Se-Cys, are locate in the 3' UTR of the corresponding pro- and eukaryotic mRNAs (Berry 2000). In prokaryotes, Se-Cys insertion in the peptide chain, is conferred by SECIS-binding proteins (SELB) and the translation machinery. Examples of eukaryotic Se-Cys containing proteins: glutathione peroxidase (Shen et al. 1993), type I iodothyroinine deiodinase (Berry et al. 1991), thioredoxin reductase (Buettner et al. 1999), selenoprotein P (Hill et al. 1993; Fig. 2.4).

-De-adenylation signals located in the 3' UTR lead to a shortening of the poly(A)-tail, followed by an mRNA decapping (Muhlrad and Parker 1992; Decker and Parker 1994, 1995; Schwartz and Parker 2000).

-AU-rich elements in the mRNA 3' UTR can be involved in determining the short half-life of transiently expressed genes (Theodorakis and Cleveland 1996). Mammalian *c-fos* mRNA contains a 75 nts AU-rich element in the 3' UTR, which promotes rapid mRNA decay. The extremely short half-life of *c-fos* mRNA is crucial for the transient expression of the *c-fos* genes and is also determined by a second instability element in the ORF (Shyu et al. 1989). A further example is a 51 nt AU-rich sequence in the 3' UTR of lymphokine granulocyte-monocyte colony stimulating factor (GM-CSF) mRNA (Shaw and Kamen 1986).

-The poly(A) tail at the 3' end of an mRNA binds the poly(A)-binding protein (PABP), which enhances cap-dependent translation. It has been shown that the poly(A) tail can promote the recruitment of the small ribosomal subunit to the mRNA. This function of the poly(A) tail is mediated by interactions between PABP and the eukaryotic initiation factor eIF4G. The simultaneous binding of eIF4G to the cap binding protein eIF4E forms a loop that can physically approximate the 3' and 5' end of an mRNA (Tarun and Sachs 1995, 1996; Jacobson 1996; Tarun et al. 1997; Wells et al. 1998; Sachs 2000)

-In the *Xenopus* Xhlbox2 mRNA 3' UTR multiple motifs (19nt consnesus), which are targets for a sequence specific endo-ribonuclease overlap with the recognition motif for a protective factor Brown et al. 1993, Decker and Parker 1994).

-The 3' UTR of mammalian transferrin-receptor (TfR) mRNA contains a complex structure consisting of five iron-response elements (IREs) to which the IRE binding-protein (IRP) binds when cellular iron is limiting. This interaction at the 3' UTR prolongs the half-life of TfR mRNA by stabilizing the transcript against specific endo-nucleolytic cleavage (Casey et al. 1988; Muellner and Kuehn 1988, Hentze and Kuehn 1996, (Fig. 2.7). An IRE-like structure in the 3' UTR mouse liver glycolate oxidase (GOX) mRNA binds IRP *in vitro*, but does not confer iron dependent regulation (Kohler et al. 1999)

-developmental signals in the mRNA 3' UTR (Wickens et al., 1997, 2000). In mRNAs coding for different proteins in organisms like *Drosophila* and *Caenorhabditis elegans* regulatory signals can be found, examples are: for *Drosophila, bicoid* mRNA (Macdonald and Struhl 1988 Fig. 2.6); the *nanos* response element in *hunchback* mRNA 3' UTR (Wharton and Struhl 1991) and several regulatory elements within the *oskar* mRNA 3' UTR (Kim et al. 1993). In *C.elegans tra2* and *-fem*3 mRNA 3' UTR, which bear translational regulatory elements *dre* and *-pme* respectively involved in germline pattering control (Ahringer and Kimble 1991; Goodwin et al. 1993; Wickens et al. 2000).

-Histone mRNAs are not polyadenylated, but end in a highly conserved 26 nts sequence,containing a 16 nt stem loop. This stem loop and the purine-rich histone downstream element (HDE) are recognized by two trans-acting factors: the stem loop binding protein (SLBP) and the U7snRNP (Mowry and Steitz 1987; Martin et al. 1997; Dominski and Marzluff 1999).

-An element in mouse protamine 2 mRNA 3' UTR temporally represses its translation by interaction with a 18 kDa protein (Kwon and Hecht 1993).

- **Motifs in the 5' UTR and 3' UTR**

-In D*rosophila msl-2* mRNA 5' UTR two consensus sex-lethal (SXL) binding sites are located in the intron, where SXL binding inhibits splicing. Four of them are positioned in the 3' UTR. Following export into the cytoplasm, SXL binding to the sites within both UTRs represses translation of *msl-2* mRNA (Gebauer et al. 1997, 1999).

- **Motifs in viral RNAs**

(Examples are also given in Figs. 2.17 and 2.18).

-Coliphage MS2 RNA (Fig. 2.17) hairpin motif at the start of the coat protein gene, (Olsthoorn et al. 1994)

-Common regulatory motifs in PP7-, Qß-, MS2- and M11-phages (Olsthoorn et al. 1995; Fig 2.17) coat initiator motif (energy delta-Go fine-tuned for maximal ribosome loading); replicase start motif; 5' core; RNA secondary structure at the conserved amino acid motif YGDD

-rev-response core element (Cullen and Malim 1991) interacting with rev protein (Fig. 2.17). A human nucleoporine interacts also specifically with HIV Rev (Fritz et al. 1995).

-TAR-RNA from HIV bound by the Tat protein (Fig. 2.17; Moras and Poterszman 1996)
-regA RNA from T4 ; motif: 5'AAUGAGGAAAUU (Draper 1995)
-R17 RNA and coat protein (Draper 1995)
-Adenovirus 2 virus-associated RNA inhibits activation of the host protein kinase DAI as defense (Ma and Mathews 1993); highly structured and conserved. motif (Fig. 2.17).
-HTLVI, HTLVII contain an element in R/U5 of the (Rex) 5' terminal long repeat: the rex-response element (Black et al. 1994; Fig. 2.17).

Antisense RNAs
-P22 phage antisense sarRNA (Fig. 2.18) supresses the antirepressor protein (Wu et al. 1987)
-*E.coli* regulatory antisense RNA *micF* inhibits *ompf* mRNA (Mizuno et al. 1984; Fig. 2.18)
-*C.elegans* lin-4 RNA inhibits *lin-14* RNA by antisense binding to 31 nts (Lee et al. 1993; Wightman et al. 1993)
-Tn10 transposase regulated by small complementary RNA (Simons and Kleckner 1983; Fig. 2.18).

Regulatory and autocatalytic motifs and structures
-The plant rhizobial RNA from the sra gene (213 nts) forms a stem-loop secondary structure (Ebeling et al. 1991)
-Hammerhead motif (Symons 1992; Long and Uhlenbeck 1993)
Simple and double hammerhead RNA motifs (Fig. 2.21)
Hairpin motif (Long and Uhlenbeck 1993; Fig. 2.22);
Delta ribozyme (Long and Uhlenbeck 1993)
-RNAse P cage structure of the RNA component (Fig. 2.23). RNAse P substrate motifs tRNA and 4.5S rRNA (Altman et al. 1993; Fig. 2.24)
-Group I introns (Cech 1993; Saldanha et al. 1993) conserved structure (Fig. 2.25)
-Group II introns (Cech 1993; Michel and Ferat 1995). Typical structure (Fig. 2.26)

-7S RNA, the RNA part of the signal recognition particle (SRP) involved in ER translocation; 7S RNA as central part, highly structured, stable secondary structure (Lütcke 1995; Fig. 2.27)
-Y-RNA (van Horn et al. 1995) putatively involved in 5S rRNA synthesis, part of Ro RNP; in all vertebrate cells, also found in *C.elegans*. Pyrimidine-rich internal loop, long stem in which the 5' and 3' site ends are base-paired. Within the stem is a conserved bulged helix that is proposed to bind the Ro-protein.

Table 2.2 gave an overview of selected important RNA regulatory motifs and structures; many more are discussed with respect to their different functions in this chapter, as well as in Chapters 5 and 6.

2.16 RNA Motifs in RNA Viridae

RNA viridae are rich in regulatory RNA structures. Often, many functional elements are present in a comparatively small stretch of RNA, enabling fine-tuned regulation of mRNA translation and stability, antisense regulation and other regulatory steps, to ensure optimum survival chances and high adaptive capacity for the RNA virus.

Table 2.3 gives a concise overview of the different types of RNA viruses. The underlying nucleotide sequences for each virus can easily be retrieved by sequence retrieval systems, including detailed comments ("features" of the sequence) on the coding region and on the regulatory RNA elements where they are known, such as LTRs in retroviridae (see Chap. 4). The secondary structure for regions and motifs of interest can be obtained using folding programs (introduced in Chap. 4).

Table 2.3. Classification of different RNA viridae

dsRNA viridae

Host	*Virus*
Bacteria	Cystoviridae
Fungi	Partitviridae, Partitvirus Chrysovirus, Totiviridae, Hypoviridae
Plants	Reoviridae, Phytoreovirus, Fijivirus, Oryzavirus, Partitviridae, Alphacryptovirus, Betacryptovirus
Invertebrates	Reovirdae, Cypovirus, Birnaviridae, Entomobirnavirus,
Vertebrates	Reoviridae, Orthovirus, Orbivirus, Coltivirus, Rotavirus, Aquareovirus, Birnaviridae, Aquabirnavirus, Avibirnavirus

ssRNA(-) strand viridae

Host	*Virus*
Plants	Bunyaviridae, Tospovirus, Tenuivirus, Rhabdoviridae, Cytorhabdovirus, Nucleo-rhabdovirus

Invertebrates	Bunyaviridae, Rhabdoviridae
Vertebrates	Filoviridae, Bunyaviridae, Bunyavirus, Hantavirus, Nairovirus, Phlebovirus, Arenaviridae, Paramyxoviridae, Bornaviridae, Rhabdoviridae, Lyssavirus, Vesiculovirus, Ephemerovirus, Orthomyxoviridae

ssRNA(+) strand viridae

Host	*Virus*
Bacteria	Leviviridae
Fungi	Barnaviridae
Plants	Enamovirus, Idaeovirus, Sequiviridae, Tombusviridae, Luteovirus, Marafivirus, Sobemovirus, Tymovirus, Comoviridae, Potexvirus, Capillovirus, Trichovirus, Carlavirus, Potyviridae, Closteroviridae, Bromoviridae, Cucumovirus, Bromovirus, Alfamovirus, Ilarvirus, Tobamovirus, Tobravirus, Hordeivirus, Furovirus
Invertebrates	Picornaviridae, Nodaviridae, Tetraviridae, Flaviviridae, Togaviridae
Vertebrates	Picornaviridae; Calciviridae, Astroviridae, Flaviviridae, Togaviridae, Arteriviridae, Coronaviridae

ssRNA-RT viridae

Host	*Virus*
Vertebrates	Retroviridae

2.17 Protein Motifs Pointing to Interacting RNA Structures

Regulatory RNA and RNA structural elements interact with different proteins in many instances (Table 2.2; further details are found in Chap. 5 and Chap. 6). However, basic principles and amino acid motifs in the interacting protein structure recur and are summarized in Table 2.4.

Table 2.4. Important protein motifs involved in RNA recognition

---

Nuclear localization signals (NLS)
-M9-domain, the C-terminal 38-residue NLS of hnRNP A1 (Michael et al. 1995)
-Bipartite basic localization signal (Robbins et al. 1991)
-hnRNP K specific nuclear shuttling domain (KNS) (Michael et al. 1997)

Arginine-rich motifs (Dandekar and Argos 1996; Harada et al. 1996)
-Tat protein from HIV
-Nucleolin (Dreyfuss et al. 1993)
-Hepatitis delta antigen (Lee et al. 1993)

RGG box (Dreyfuss et al. 1993)
-An RNA binding domain consisting of RGG (Arg-Gly-Gly) repeats, first identified in the C-terminus of hnRNP U by deletion mapping (Kiledjian and Dreyfuss 1993). In RGG boxes, RGG repeats are often interspersed with aromatic amino acids Dreyfuss et al. 1993). In hnRNP A2, RGG repeats have been shown to be critical for the cellular localization of the protein (Nichols et al. 2000).

KH domain
-Regulatory RNA binding motiv, first described as hnRNP K homology domain (KH domain). A whole family of proteins, which is still growing, bearing this motif has been identified (examples: vigilin, FMR1, hnRNPs K, E1, E2). A NMR structure has become available for a KH domain of vigilin (Musco et al. 1996).

ZINC-finger
ZINC-fingers have been identified in TFIIIA which binds 5S RNA and human nucleoporin (Dreyfuss et al. 1993; Fritz et al. 1995)
Other RNA recognition motifs

-RNA recognition motif in *Drosophila rox8* (Brand and Bourbon 1993)
-poly(A) binding protein (PABP) (Sachs and Kornberg 1985).
-    RNP 1, RNP2 (Ripmaster and Woolford 1993).

2.18 Modified Nucleotides

Besides the standard nucleotides, RNA can also contain modified nucleotides, which are recognized by other molecules in the cell. They represent micromotifs, which are essential for specific functions of the RNA molecule, e.g. regulation snRNA import and export and mRNA export (see Chap. 5) or ribosomal RNA (see Chap. 6). Sometimes the reason for the modification is not known. The following Table 2.5 lists the more important examples. Such modifications normally can only be revealed by direct probing and are often ignored by sequence analysis and gene cloning. Taking account of inosine and other modified nucleotides with additional base pairing capacities may greatly enhance sensitivity of motif searches (Chap. 4) and is also important in experimentally delineating base pairings (e.g. in mutagenesis, Chap. 3).

Table 2.5. RNA modification

mRNA
-m7GpppN-cap, the 5' terminal nucleotide in cellular RNAs (Shatkin 1976)
-internal 6-methyladenine residues with a strict consensus sequence Gm_6AC or Am_6AC (Tuck 1992).

tRNA
Pseudo uridylation: use of non-canonical nucleotides for instance inosine in the third position of the anticodon (counted from 5') may pair with adenosine, uracile and cytosine (the wobble base). There are several more modifications, some quite intricate, regulating codon fidelity. Usually they are at position 37, the first position past the anticodon (Gesteland and Atkins 1993).
rRNA

Often guide RNAs are used to insert specific modifications (Kiss-László et al. 1996).
-Ribose methylation is the most prominent modification.

-Pseudo uridylation
-Adenosin-methylation by ErmE methyltransferase, a modification found in *Saccharopolyspora erythraea* (Villsen et al. 1999).

snRNAs
Modified in a way similar to rRNA (e.g. U4/U6), including guide RNAs for positioning in ribose-methylation.

2.19 Perspective: from Known Instances to New Ones

This chapter illustrated the characterized processes in the cell where RNA is a key player in cellular metabolism (summarized in Table 2.1) and utilizes different motifs, i.e. short segments of the whole RNA molecule, for specific interaction steps. The specific interaction of an RNA motif with other RNAs or proteins exerts in most cases further control on the reaction in which the RNA molecule containing the motif is involved, and can thus often be described as a regulatory motif. An overview of the known instances of RNA motifs is illustrated in the text and figures of this chapter and catalogued in Table 2.2. The overview is provided as a basis in the search for further RNA molecules. Especially motif-rich are RNA viridae (Table 2.3). RNA recognition motifs in proteins indicate that there is an interacting RNA partner molecule, which should be identified (Table 2.4). Finally, a potential complication to keep in mind is modified nucleotides (Table 2.5).

2.20 An Atlas of Regulatory RNA Elements

In the following we skech the most important functional features of common regulatory RNA elements and structures.

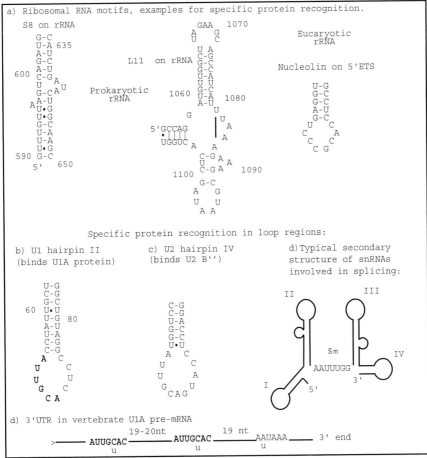

a) Ribosomal RNA motifs, examples for specific protein recognition.

Specific protein recognition in loop regions:

b) U1 hairpin II
(binds U1A protein)

c) U2 hairpin IV
(binds U2 B'')

d) Typical secondary
structure of snRNAs
involved in splicing:

d) 3'UTR in vertebrate U1A pre-mRNA

Fig. 2.2a Shown are binding sites on prokaryotic rRNA (Draper 1995; S8 binding site (Kalurachchi et al. 1997) and L11 binding site (Guttel et al. 1994) in *E.coli* rRNA) and of the protein nucleolin on the 5' external transcribed spacer of eucaryotic rRNA. The nucleolin stem loop requires the sequence UCCCGAA in the loop region and has also been confirmed by SELEX experiments. (Ghisolfi-Nieto et al., 1996) **b,c,d,e** Specific protein interactions with small nuclear RNAs.
Drawn are the schematic secondary structures of U1 RNA which interacts with U1A protein (**b**) and U2 interaction with U2 B'' protein (**c**). Interaction of U1 with U1A is specific, whereas, B'' requires additional binding of U2A'protein (Scherly et al. 1990) and the interaction of the 3' UTR of U1A pre-mRNA with U1A protein (Boelens et al. 1993). A sketch of the typical snRNA secondary structure including the Sm protein binding site (nucleotides) and four major stem loops are shown in **d**. The U1A protein-binding loop in snRNA U1 pre-mRNAs from human, mouse and *Xenopus* display the same motif as shown in **e**. The conserved nucleotides are drawn in bold and the variants in small caps; the whole mRNA area is 85% identical in all three species.The polyadenylation motif has the variant AUUAAA in mouse and humans

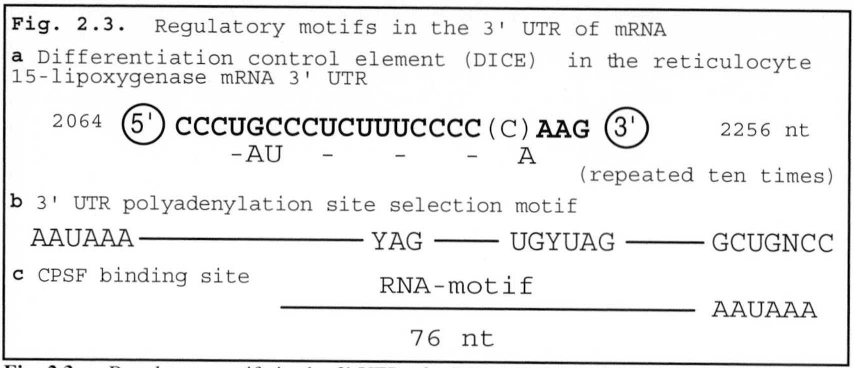

Fig. 2.3.   Regulatory motifs in the 3' UTR of mRNA

a Differentiation control element (DICE) in the reticulocyte
15-lipoxygenase mRNA 3' UTR

2064   (5') **CCCUGCCCUCUUUCCCC** (C) **AAG** (3')   2256 nt
　　　　　-AU -   -   - A
　　　　　　　　　　　　　　　　　　(repeated ten times)

b 3' UTR polyadenylation site selection motif

AAUAAA ——————————— YAG —— UGYUAG ——— GCUGNCC

c CPSF binding site   RNA-motif
　　　　　　　　　　　——————————————————— AAUAAA
　　　　　　　　　　　　　　76 nt

Fig. 2.3a-c Regulatory motifs in the 3' UTR of mRNA. **a** The rabbit LOX 3' UTR repeat (15-lipoxygenase mRNA 3'UTR repeat. Ostareck-Lederer et al. 1994). The consensus sequence is shown in *bold* and deviations from the consensus are indicated. Translation of DICE-bearing mRNAs is silenced by binding of hnRNPs K and E_1 to the DICE. The hnRNPs K/E_1 complex inhibits 60S ribosomal subunit joining to the 40S subunit recruited the the AUG. **b** An example from several identified motifs in mammals, implicated in polyadenylation site selection by binding to CstF (cleavage-stimulating factor; Keller 1995). The distance between the tri-partite motifs is maximally five nucleotides, whereas the distance between the polyadenylation site and the motif can be up to 50 nt in length. **c** Schematic diagramm of the CPSF (cleavage and polyadenylation-stimulating factor) binding site. The 76 nt upstream of AAUAAA are required in HIV for efficient polyadenylation by the polyadenylation specificity factor CPSF (Gilmartin et al.1995). However, the nature of the exact RNA motif is not yet known

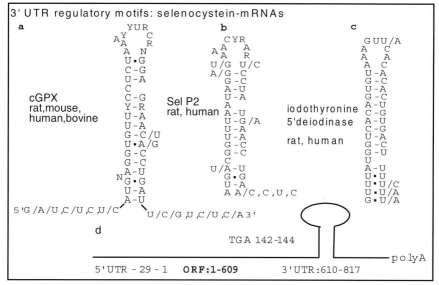

Fig.2.4a-d. 3' UTR regulatory structures in selenocysteine-mRNAs (Se-Cystein insertion elements (SECIS)). The examples illustrated are: **a** Glutathione peroxidase. Gluthathione peroxidase is required to incorporate selene via selenocysteine for redox-protection.**b** Selenoprotein P is another cellular protein incoporating selenium. **c** Type I iodothyronine 5' deiodinase also utilizes selenium for its catalytic activity (Berry et al. 1991). The regulatory structure induces read over of the UGA stop codon and decoding UGA for Se-Cys leading to the incorporation of the selenocysteine (Shen et al. 1993). **d** A schematic diagramm indicating the positions of the ORF, the 3' UTR and the selen encoding TGA codon within the human cellular glutathione peroxidase is shown.The 3' UTR cGPX motif (see **a**) shows only part of the structure for simplicity; a further stem loop structure in the ORF between nt 143 and 210 is not shown

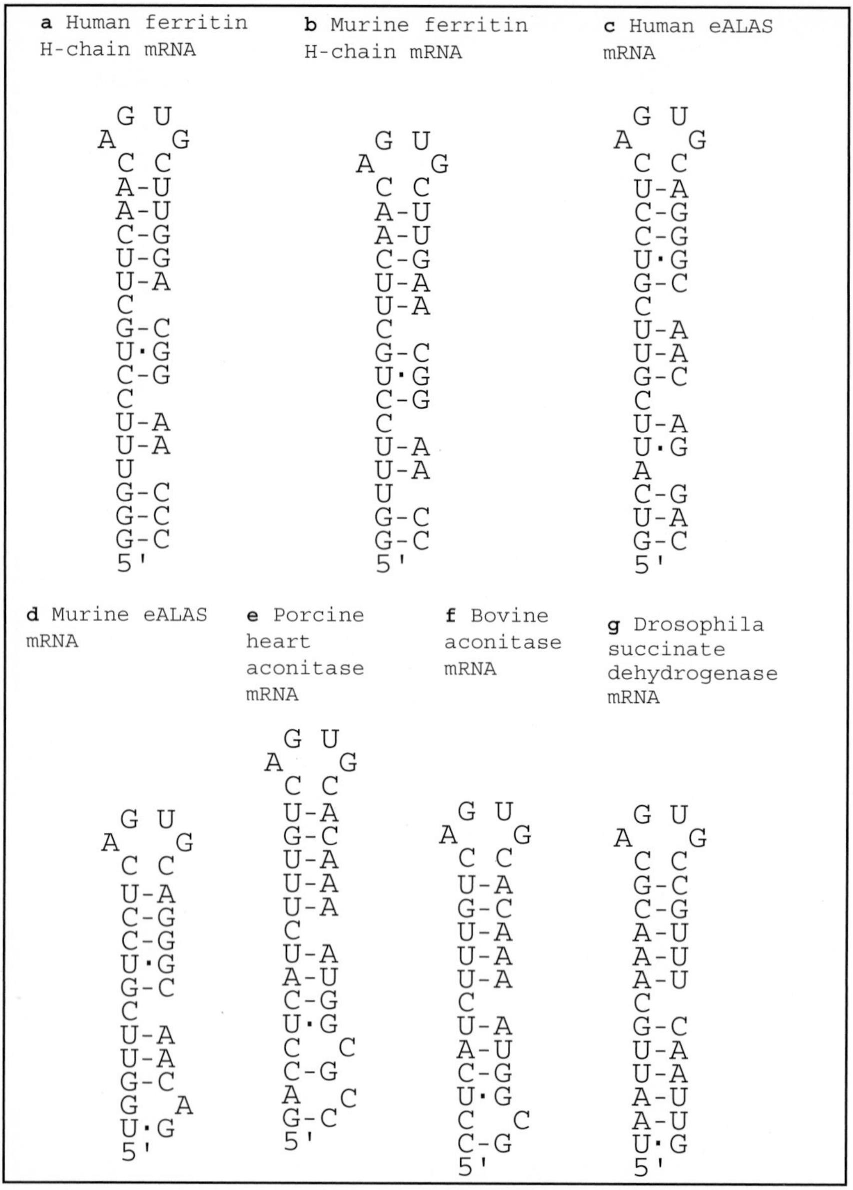

a Human ferritin
H-chain mRNA

b Murine ferritin
H-chain mRNA

c Human eALAS
mRNA

```
     G U
   A     G
    C   C
    A-U
    A-U
    C-G
    U-G
    U-A
    C
    G-C
    U·G
    C-G
    C
    U-A
    U-A
    U
    G-C
    G-C
    G-C
    5'
```

```
       G U
     A     G
      C   C
      A-U
      A-U
      C-G
      U-A
      U-A
      C
      G-C
      U·G
      C-G
      C
      U-A
      U-A
      U
      G-C
      G-C
      5'
```

```
     G U
   A     G
    C   C
    U-A
    C-G
    C-G
    U·G
    G-C
    C
    U-A
    U-A
    G-C
    C
    U-A
    U·G
    A
    C-G
    U-A
    G-C
    5'
```

d Murine eALAS
mRNA

e Porcine
heart
aconitase
mRNA

f Bovine
aconitase
mRNA

g Drosophila
succinate
dehydrogenase
mRNA

```
     G U
   A     G
    C   C
    U-A
    C-G
    C-G
    U·G
    G-C
    C
    U-A
    U-A
    G-C
    G   A
    U·G
    5'
```

```
       G U
     A     G
      C   C
      U-A
      G-C
      U-A
      U-A
      U-A
      C
      U-A
      A-U
      C-G
      U·G
      C   C
      C-G
      A   C
      G-C
      5'
```

```
     G U
   A     G
    C   C
    U-A
    G-C
    U-A
    U-A
    U-A
    C
    U-A
    A-U
    C-G
    U·G
    C   C
    C-G
    5'
```

```
     G U
   A     G
    C   C
    G-C
    C-G
    A-U
    A-U
    A-U
    C
    G-C
    U-A
    U-A
    A-U
    A-U
    U·G
    5'
```

Fig. 2.5a-g Iron-responsive elements in the 5' UTR from different mRNAs.
a human ferritin H-chain mRNA. **b** murine ferritin H-chain mRNA. **c** human eALAs mRNA **d**
murine eALAS mRNA **e** porcine mitochondrial heart aconitase mRNA. **f** bovine mitochondrial
aconitase mRNA (see Dandekar et al. 1991 and Gray et al. 1996 for details on the particular
IREs)

```
Developmental regulatory elements in the 3' UTR of mRNA

a) nanos response elements in the 3' UTR of hb (hunchback):
   GUCCCCAUCACCUUGUUAUUAUUAUUAUUAUCACUAUUAU

   CAUAUAAUCGUUGUCCAGAAUUGUAUAUAUUCGUAGCAUAAGUUUUCCAAAC hb1

   AUUAUUUUGUUGUCGAAAAUUGUACAUAAGCCAAUUAAGCCGCUAAUUCA   hb2

b) nanos response elements in the 3' UTR of bcd (bicoid):
   AACCACUGUUGUUCCUGAUUGUACAAAUACCAAGUGAUUGUAGAU      bcd
```

Fig. 2.6a,b. Developmental regulatory elements in the 3' UTR of mRNA.
Different nanos response elements are shown. **a** 145 bp of hb (hunchback) mRNA and **b** 45 bp
of bcd (bicoid) mRNA that contain nanos response elements (for their functional definition see
Wharton and Struhl 1991). The core motif is shown in *bold*, this is repeated twice in the 3' UTR
of hb and once in the 3' UTR of bcd.

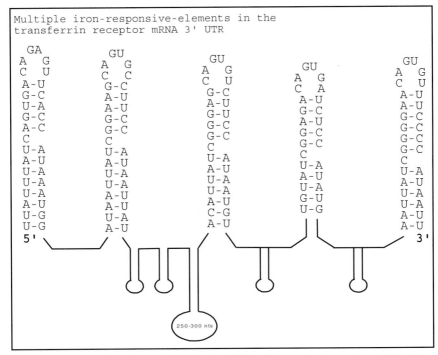

Multiple iron-responsive-elements in the
transferrin receptor mRNA 3' UTR

Fig. 2.7. Iron-responsive elements (IREs) in the 3' UTR of the transferrin receptor-mRNA (Tfr).
Shown are the human IRE sequences (Müller and Kühn 1988) in the region encompassing
nucleotides 3200 - 3800 of the transferrin receptor mRNA; chicken Tfr has a very similar
secondary structure. Further stem loops in the 3'UTR are indicated at the bottom; these are not
well conserved and no IREs are present in these regions

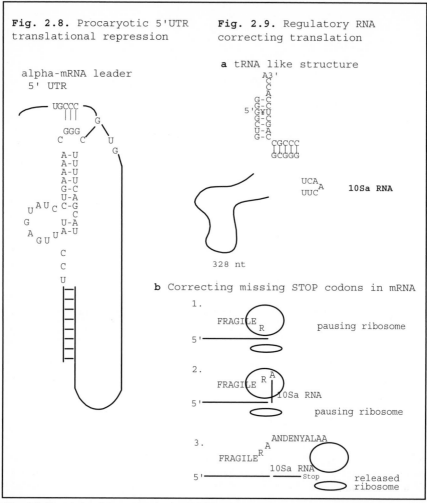

Fig. 2.8. Procaryotic 5'UTR translational repression

Fig. 2.9. Regulatory RNA correcting translation

a tRNA like structure

328 nt

b Correcting missing STOP codons in mRNA

1.

FRAGILE R   pausing ribosome

2.

FRAGILE R A   10Sa RNA

pausing ribosome

3.

ANDENYALAA
FRAGILE R A
10Sa RNA   Stop   released ribosome

Fig. 2.8. Translational repression in the 5' UTR of prokaryotes. The 5' leader sequence of the alpha-mRNA from *E.coli* is shown, The alpha-mRNA encoded ribosomal protein S4 acts as its own translational repressor by binding to the regulatory RNA structure in the leader (Deckman and Draper 1987). Secondary structures and a possible pseudoknot structure in the leader relevant for S4 protein binding are drawn

Fig. 2.9ab. Regulatory RNA correcting translation.
Shown is the 10Sa RNA from E.coli **a** Highly conserved tRNA-like structure (only minor differences in different prokaryotes; Komine et al. 1994). **b** Schematic diagramm showing translation of an mRNA by a ribosome. The missing STOP codons are corrected by 10Sa RNA. *1* mRNA with missing STOP codon leads to synthesis of the peptide FRAGILER but then the ribosome is stalled and not released. *2* In this state it is possible for 10Sa RNA to dock; 10Sa RNA mimics a tRNA charged with alanine; *3* After attaching the alanine the ribosome switches to 10Sa RNA and translates the remaining peptide encoded in 10Sa RNA until it is released by the STOP codon provided by the 10Sa RNA. The tag ANDENYALAA leads to rapid degradation of the peptide encoded by the mutilated messenger RNA

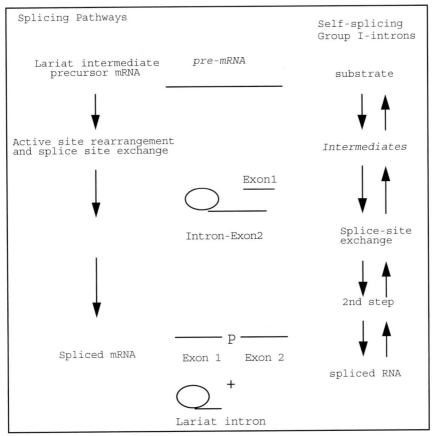

Fig. 2.10a. Schematic figure illustrating splicing pathways. Compared is the pathway of mRNA splicing and the splicing pathway in self-splicing introns (group I). Splicing of group II intron leads to a branched intermediate and lariat using an adenosine at the branch point whereas a group I intron uses a catalytic guanosine for attack and there is no lariat intermediate. Exons,the lariat intermediate and a central phosphate bond are shown. The main differences are the reversibility and additional intermediates in the self-splicing intron pathway

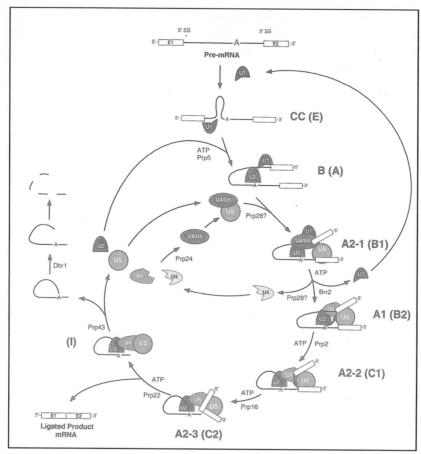

Fig. 2.10b Schematic presentation of the pre-mRNA processing in the spliceosome. The initial pre-mRNA substrate (*top*) contains two exons and one intron. As the final product of the splicing process the exons are ligated (*lower left*) and the intron is, as a lariate, excised and degraded after debranching (*middle left*). Small nuclear ribonucleoprotein particles (snRNPs) are involved at distinct steps of spliceosome formation and catalysis (*shaded*). Macromolecular complexes that have been distinguished biochemically and genetically are labelled using the designation suggested for yeast with the mammalian designation shown in parentheses: CC (E), B (A), A2-1 (B1), A1 (B2), A2-2 (C1), A2-3 (C2) and (I). *Arrows* represent transitions between complexes and/or recycling of snRNP components. The debranching enzyme Dbr1 and several yeast RNA helicase motif-containing proteins implicated in the various ATP-dependent conformational changes (Brr2, Prp2, Prp5, Prp16, Prp22, Prp24 and Prp42) are also indicated. Other non-snRNP factors are omitted (After Burge et al. 1999)

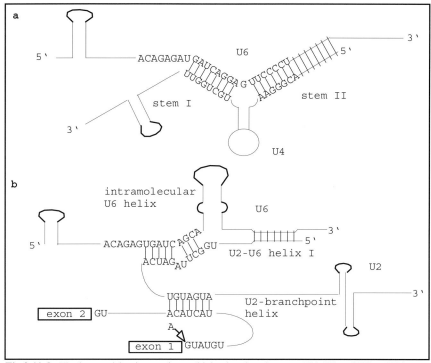

Fig 2.11ab. The base-pairing interactions which take place between U6, U4 and U2 snRNAs as a result of dynamic rearrangments during spliceosome formation. *Lines* represent single-stranded regions. **a** U4/U6 (Y-Model, Brow and Guthrie 1988; Madhani and Guthrie 1994); **b** U2/U6 interaction (Madhani and Guthrie 1994)

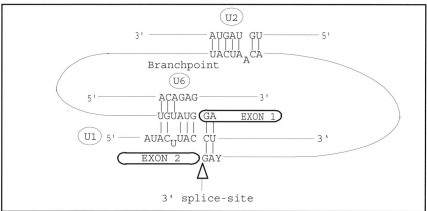

Fig 2.12. Schematic diagramm showing the base-pairing interactions of U2, U6 and U1 snRNAs with intronic sequences of the pre-mRNA. Interactions between the U2 snRNA and the branchpoint, between U6 and the 5' splice site and U1 and the 5' splice site are depicted. The selected consensus sequences of both the splice sites and the branchpoint are those of *S.cerevisiae*. For reviews on mammalian splicing interactions of these snRNAs see Green (1991), Madhani and Guthrie (1994); for yeast interactions see Rymond and Rosbash (1992)

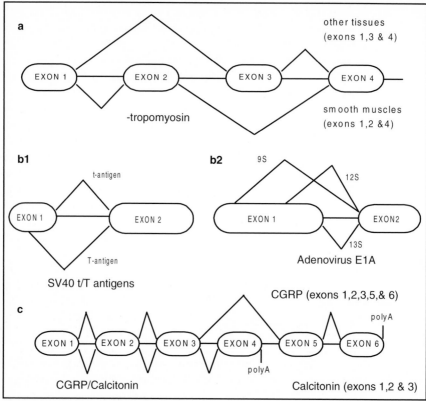

Fig. 2.13a-c Schematic figure depicting various strategies of alternative RNA splicing.
a Exclusive exon splicing of rat a-tropomyosin transcripts leads to the ligation of exons 1,3 and 4 in non-smooth muscle tissue and ligation of exons 1,2 and 4 in smooth muscles. **b1** and **b2** Selection of alternative 5' splice sites leads to the production of t and T antigens in the SV40 virus and different adenovirus E1A transcripts. **c** CPRG/caltitonin transcripts contain multiple polyadenylation sites, which are differentially selected. In neuronal cells exons 1,2,3,5 and 6 are spliced together and polyadenylated at the 3' end of exon 6, leading to the production of CPRG (calcitonin gene-related peptide), whereas in non-neuronal cells, calcitonin is produced by the exons 1,2,3 and 4 which are joined and polyadenlated at the 3' end of exon 4. Examples according to Hodges and Bernstein (1994)

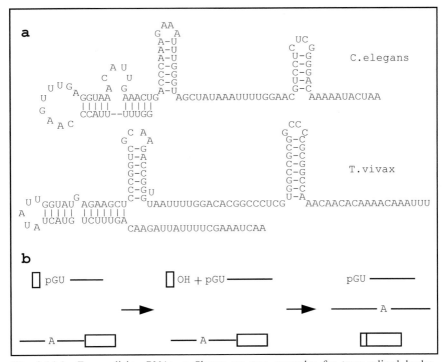

Fig. 2.14ab. *Trans*-splicing RNAs. **a** Shown are two examples for *trans*-spliced leader structures from *C.elegans* and trypanosomes **b** principle of the reaction: the leader ☐ is spliced to the exon of another message ☐▭, and a branched intermediate is released

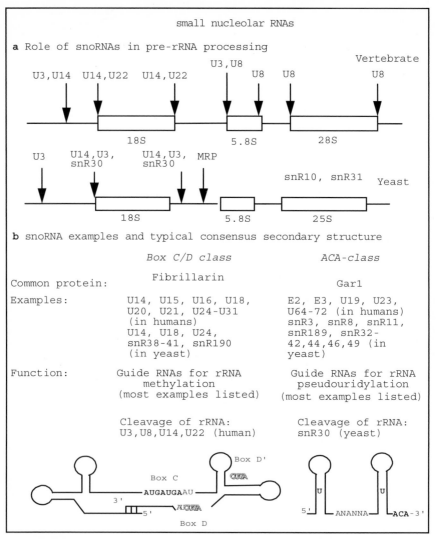

Fig.2.15ab. Small nucleolar RNAs. The different types are listed and several examples given (see Smith and Steitz 1997 for latest examples). **a** Different snoRNAs influencing cleavage steps during pre-ribosomal RNA processing are indicated (Maxwell and Fournier 1995).In yeast, snRNP proteins NOP1,GAR1 and SOF1 are required for 18S synthesis; POP1 and SNM1 are components of RNAse MRP and are required for processing of the 5' end of 5.8S.The exact role of snR10 and snR31 in processing is unclear; they are probably required for early processing steps (Dandekar and Tollervey 1993). **b** Examples and typical consensus secondary structure of the two classes of snoRNAs. Listed are the two major classes of snoRNAs, snoRNAs provide also guide RNAs for modification of other RNAs (e.g. modification of snRNA U5). Drawn in the figure are the consensus for *Box C* and *Box D* containing snoRNA (*left*). Additionaly present is the internal box D'. The 3' and 5' end of these snoRNAs share a region of complementarity; on the right a typical ACA class snoRNA is shown. *U* indicates the region of complementarity to rRNA required for its pseudouridylation

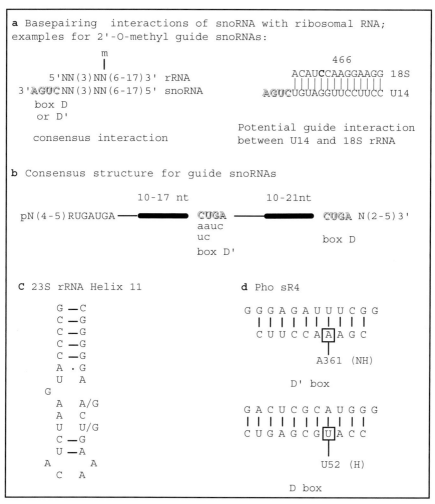

```
a Basepairing  interactions of snoRNA with ribosomal RNA;
examples for 2'-O-methyl guide snoRNAs:

             m
             |                               466
   5'NN(3)NN(6-17)3' rRNA           ACAUCCAAGGAAGG 18S
3'AGUCNN(3)NN(6-17)5' snoRNA        ||||||||||||||
                                   AGUCUGUAGGUUCCUUCC U14
   box D
   or D'
                               Potential guide interaction
   consensus interaction       between U14 and 18S rRNA


b Consensus structure for guide snoRNAs

                      10-17 nt           10-21nt

 pN(4-5)RUGAUGA ━━━━━■━━━━   CUGA ━━━━━■━━━━  CUGA N(2-5)3'
                            aauc
                            uc                   box D
                            box D'


 C 23S rRNA Helix 11           d Pho sR4

        G — C
        C — G              G G G A G A U U U C G G
        C — G              | | | | | | | | | |
        C — G              C U U C C A A A G C
        C — G
        A · G                      A361 (NH)
        U   A
     G                         D' box
        A   A/G
        A   C                G A C U C G C A U G G G
        U   U/G              | | | | | | | | | | |
        C — G                C U G A G C G U A C C
        U — A
     A       A                     U52 (H)
        C   A
                               D box
```

Fig. 2.16a-d. Small nuleolar guide RNAs. **a** rRNA undergoes extensive modifications during its maturation pathway.One such modification is methylation, which requires base-pairing between the snoRNA and rRNA (Kiss-Laszlo et al. 1996) as illustrated in a subset of examples above; the methylated rRNA nucleotide is shown *in bold*. The rRNA strand is drawn 5' to 3', the complementary snoRNA below runs in the opposite direction, i.e. 3' to 5' (consensus 3' motif box D or D' is *shadowed*). **b** Schematic drawing showing the consensus structure of guide snoRNAs. Complementarity regions to rRNA are indicated by a *thick line*. Only U14, U45, U50 and U24 snoRNA contain two regions which are complementary to rRNA. **c** For comparison the central common motif that organizes the structure of multihelix loops in 16S and 23S ribosomal RNAs is shown. (Leontis et al. 1998). **d** Recent similar interaction (sRNA/vRNA) in Archaea (Omer et al. 2000)

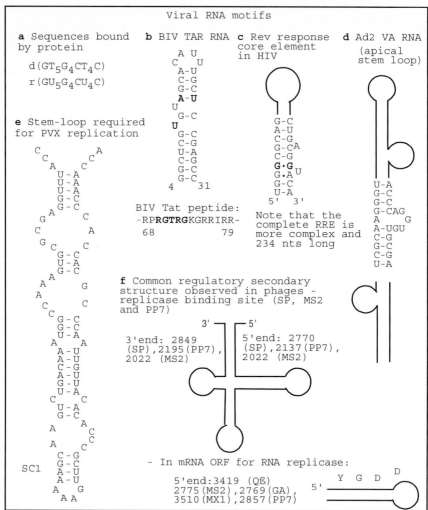

Fig. 2.17a-f. Typical regulatory motifs found in viral RNAs. **a** Single-stranded nucleic acid sequences that are bound by the gene 5 protein of filamentous bacteriophage fd. The RNA sequence is important for translational repression (Oliver et al. 2000). **b** Bovine immuno deficiency virus TAR RNA (a triple nucleotide interaction between U10,A13 and U24 is shown *in bold*); additionally shown is the TAT peptide (key residues in *bold*; Moras and Poterszman, 1996); **c** RRE core element from HIV (Dandekar and Koch 1996; Bartel et al. 1991); note a central non-canonical G·G pair in bold and another non canonical A•G pair. **d** Regulatory RNA Ad2 from adenovirus (Ma and Mathews 1993; see text). **e** A stem-loop structure is required for potato virus X replication. The figure shows the structure of the SC1 mutant. (Miller et al. 1998) **f** Common three hairpin pattern suggested to be the replicase binding site in MS2 and PP7 phages and established for SP and Qß; boundaries of the 3' and 5' ends in the different phages are indicated; also shown is the common single hairpin found in the mRNA encoding the RNA-dependent RNA replicase, found around the codons for the conserved amino acids Y,G,D and D in five different phages (Further examples including 3' UTR motifs can be found in Olsthoorn et al. 1995)

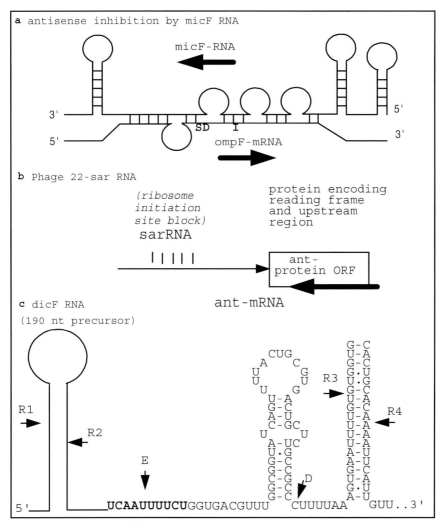

Fig. 2.18a-c. Antisense RNAs. **a** micF-RNA inhibits translation of the ompF operon. The secondary structure and orientation of both RNAs are shown. Additionally positions of the Shine-dalgarno sequence (**SD**) and initiation site (**I**) on ompF mRNA are also indicated. (Mizuno et al. 1984). **b** Schematic figure illustrating the sar RNA of the P22 phage, which blocks the expression of the ant (phage anti-repressor) protein by base pairing to its mRNA.
c DicF RNA. DicF RNA is a *trans*-acting RNA which inhibits cell division and is processed from a long precursor by multiple cleavages. (Faubladier et al. 1990) The final regulatory RNA (extending from site D to E) is generated by prior cleavages at sites D, R1 to R4 (mediated by RNAse III) and cleavage E (mediated by RNase E, shown *in bold*).

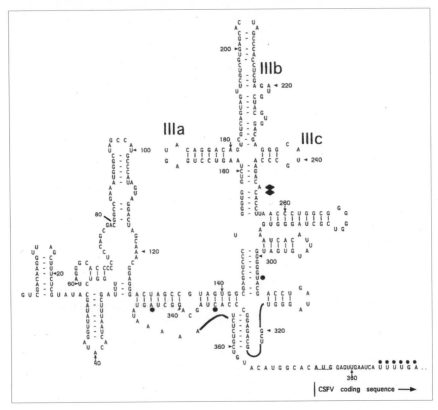

Fig. 2.19a Classical Swine Fever Virus IRES. Model of the secondary structure of the 5'UTR of Classical Swine Fever Virus (CSFV) bearing an Internal Ribosome Entry Site (IRES) (Honda et al., 1996, RNA (2), 955-968). The CSFV-IRES initiation codon is *underlined*. Stop sites were primer extension arrest was caused or enhanced by binding of eIF3 and of 40S subunits are indicated by *filled diamonds* and *circles*, respectively. A 43S preinitiation complex binds at the CSFV-IRES without initiation factors eIF4A, eIF4B, eIF4E and eIF4G. eIF3 is absolutely required for the formation of the 43S preinitiation complex, as well as the joining of the ribosomal 60S subunit and the formation of a functional 80S ribosome (After: Pestova et al., 1998)

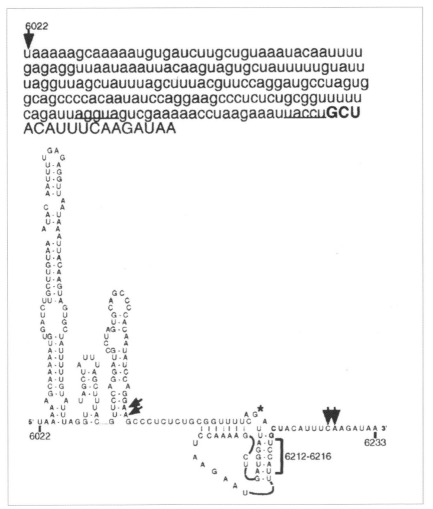

6022

uaaaaagcaaaaaugugaucuugcuguaaauacaauuuu
gagagguuaauaaauuacaaguagugcuauuuuuguauu
uaggguuagcuauuuagcuuuacguuccaggaugccuagug
gcagccccacaauauccaggaagcccucucugcgguuuuu
cagauuagguagucgaaaaaccuaagaaauuaccu**GCU
ACAUUUCAAGAUAA**

Fig. 2.19b Cricket Paralysis Virus IGR-IRES.(A) Nucleotide sequence of the intergenic region (IGR) of the dicistronic Cricket Paralysis Virus (CrPV) RNA which separates the upstream open reading frame (orf) encoding viral nonstructural genes from the downstream orf encoding the structural proteins. The sequence starts with the stop codon (UAA, nts 6022-6024) of the upstream orf, the first nucleotides of the coding sequence of the downstrem orf are indicated in *uppercase* letters, inverted repeats are *underlined* and the predicted A-site GCU triplet of the viral downstream orf is marked in *bold*. (B) The predicted folding of the CrPV-IGR secondary structure, which has been shown to contain an IRES (Wilson et al., 2000a, Mol. Cell Biol. (20), 4990-4999). The inverted repeat sequences are indicated with *brackets* and the A-site GCU triplet at which internal initiation occurs is marked in *bold*. The *arrows* indicate the position of the primary toeprints observed upon 40S ribosomal subunit binding, and the *asterix* that of an additional toeprint seen in the presence of both, 40S and 60S ribosomal subunits (Wilson et al., 2000b, Cell (102), 511-520). 80S monosomes assemble at the CrPV-IGR-IRES without the requirement of initiation factors. This unusual type of internal initiation of translation starts from the A-site of the ribosome with an alanine as the first amino acid encoded by the GCU triplet (After: Wilson et al., 2000).

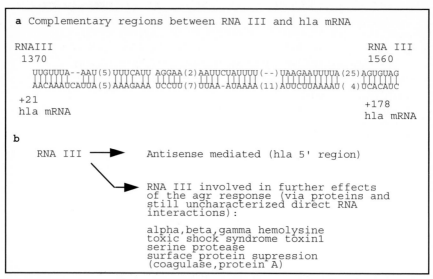

a Complementary regions between RNA III and hla mRNA

```
RNAIII                                                    RNA III
 1370                                                      1560
   UUGUUUA--AAU(5)UUUCAUU AGGAA(2)AAUUCUAUUUU(--)UAAGAAUUUUA(25)AGUGUAG
   ||||||| ||| ||||||| |||| |||||||||| |||||||||| ||||||
   AACAAAUCAUUA(5)AAAGAAA UCCUU(7)UUAA-AUAAAA(11)AUUCUUAAAAU( 4)UCACAUC
 +21                                                      +178
 hla mRNA                                                 hla mRNA
```

b

RNA III ⟶ Antisense mediated (hla 5' region)

RNA III involved in further effects
of the agr response (via proteins and
still uncharacterized direct RNA
interactions):

alpha,beta,gamma hemolysine
toxic shock syndrome toxin1
serine protease
surface protein supression
(coagulase,protein A)

Fig. 2.20a,b. RNAIII. This RNA (Novick et al. 1993) controls several virulence factors (agr locus in *Staphylococcus aureus*) on the transcriptional level by means of intermediary protein factors, but also by direct interaction with target gene transcripts (antisense regions). An antisense example is shown in **a** the translational start of hla mRNA is at +433 encoding alpha-hemolysine. **b** further effects of the agr response at the end of exponential growth

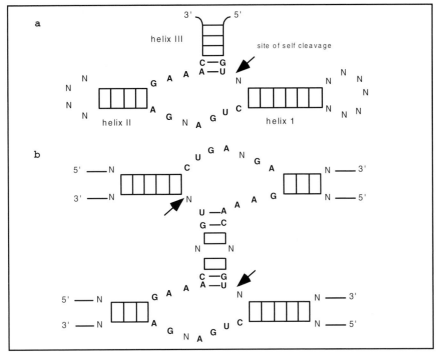

Figure 2.21a,b. a Schematic figure showing hammerhead structures present in (+) strands of different viruses. Conserved nucleotides are drawn *in bold*. Base-paired regions are drawn as boxes with *vertical bars*. N indicates that any nucleotide can be present at this position. (Long and Uhlenbeck 1993). **b** Stable, double hammerhead structure for instance found in (+) ASBV Virus. *Arrows* indicate the self-cleavage sites. See Symons (1992), Hertel et al. (1996) and Long and Uhlenbeck (1993) for detailed examples

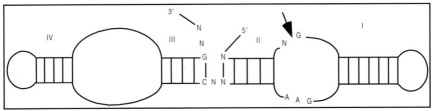

Figure 2.22. Schematic secondary structure of the hairpin ribozyme. The *arrow* indicates the self-cleavage site; any of the four nucleotides can be present at the positions marked *N*. Double stranded helical regions are drawn as *boxes with vertical bars*. *Roman letters* denote the helices. Important nucleotides according to Long and Uhlenbeck (1993) are indicated. (Long and Uhlenbeck 1993).

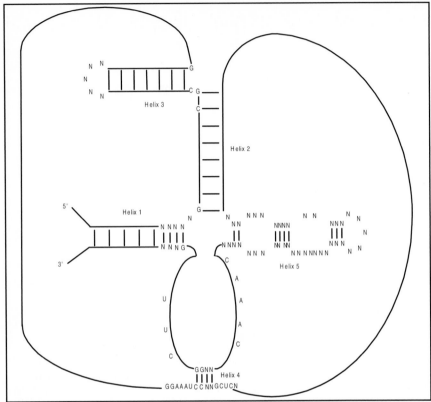

Fig 2.23. Schematic diagramm depicting the predicted cage structure of RNAse P from eubacteria. Double-stranded helical regions are numbered and drawn as boxes. Only the conserved nucleotides are shown, approximate size of the cage tertiary structure is given. Specific examples are in Gesteland and Atkins (1993) and Altman et al. (1993).

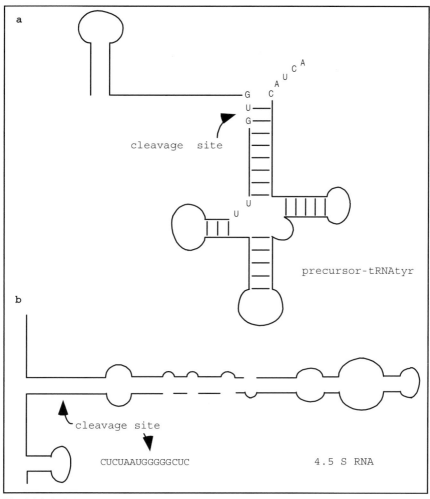

Fig. 2.24a,b. Schematic diagramm showing the secondary structures of two subsrates which are cleaved by RNAse P. **a** *E.coli* tRNA$^{tyr}_{su3}$ precursor. **b** Precursor of 4.5S RNA. Specific features of primary sequence and secondary structure are highlighted. For retrieval and folding of the RNA sequences see Chapter 4

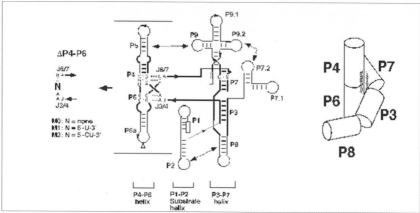

Fig. 2.25 The structure of the conserved core region of group I intron ribozymes. **a** Secondary structure of the T4 td group I intron ribozyme and its DP4-P6 derivates. *Arrowheads* superimposed on lines indicate 5' to 3' exons are indicated as *thick gray bars*. The universally conserved core region consisting of two helical domains (P4-P6 and P3-P7) is indicated by *thick lines*. The nucleotides forming base triples joining P4-P6 and P3-P7 together are also shown. Peripheral regions not directly involved in the mechanism of catalysis are indicated by *thin lines*. Tertiary interactions between peripheral regions are indicated as *broken lines with two arrowheads*. **b** A schematic three dimensional structure of the conserved core of group I intron ribozymes. The *gray shaded area* is a molecular cleft generated by specific stacking of the two helical domains in which the catalytic site is suggested to be located. (After: Ikawa et al. 2000)

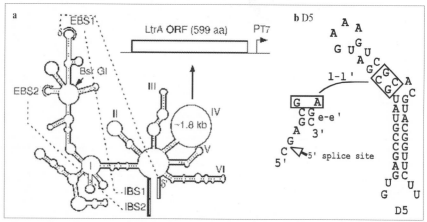

Fig 2.26a Group II intron. Schematic of the Ll.LtrB intron showing base-pairing interactions EBS1-IBS1, EBS2-IBS2, and d-d' between the intron and flanking exons. The inset shows the location of the LtrA ORF and the T7 promoter introduced into intron domain IV in donor plasmids. (After: Guo et al. 2000). **b** A set of tertiary interactions (l-l') links catalytically essential regions of D5 and D1, creating the framework for an active-site and anchoring it at the 5' splice site. Highly conserved elements similar l-l'interaction are found in the eucaryotic spliceosome. The l-l' interaction involves the fifth intronic nucleotide, which is invariably a guanine in group II introns and most in nuclear introns. In spliceosomal introns, G5 is part of a three-base-pair helix between the first serveral nucleotides of the intron and a section of U6 RNA called the 'ACAGA-box', forming a duplex that may be analogous to the e-e' interaction (Boudvillain et al.2000).

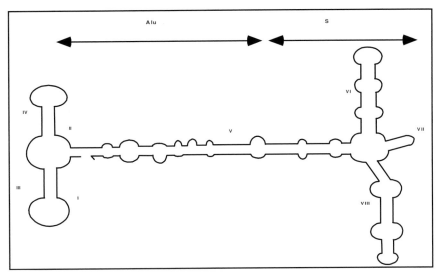

Figure 2.27. Schematic diagram depicting the secondary structure of SRP RNA from
H. sapiens. Base-paired regions are drawn as parallel lines; single-stranded regions are indicated
as *bulges*. The alu and S regions are shown as established by micrococcal nuclease digests.
(Gundelfiger et al. 1983) Helices are labelled with *Roman numbers* (I-VIII) (see Lütcke 1995 for
review)

3. Experimental Identification of New Functional RNA

3.1 Introduction

Depending on attitude and experience, both computational and experimental approaches are viable routes for revealing new RNA containing functional RNA structures. The experimental approaches are sometimes less direct but are more flexible and may lead to unexpected discoveries, for instance a regulatory protein which is even more important than the regulatory RNA in the system studied. The theoretical approaches are more focused; however, they are somewhat more biased as they require some preconception of at least the basic features of the RNA structure before the advantages of a systematic and direct search for suitable regulatory RNAs can be exploited. Furthermore, the versatile theoretical approach of a consensus search is limited by the supposition that the RNA molecule is hidden somewhere in DNA or RNA already available as a sequence. However, other theoretical and computational analyses do not necessarily require a template or even a known RNA example previously characterized to succeed (Chap. 4).

RNA motifs are both recognized by proteins and involved in RNA-RNA interactions. The RNA motif provides an important clue to the identity of the interacting partner. If this is an RNA, this often involves direct base-pairing interactions. Thus, sequencing different RNAs required in a process may lead to the discovery of potential base complementarity between them, such as between U1 and the pre-mRNA splice site. A stronger experimental verification for such postulated interactions is compensatory mutations. Such studies have been conducted on several snRNA interactions, for instance on the interaction between U4 and U6 snRNA (Madhani and Guthrie 1994; Raghunathan and Guthrie 1998). A similar direct RNA-RNA interaction is involved in pre-rRNA processing and small nucleolar RNAs. An even more direct experimental proof of the functional RNA-RNA interaction can be attempted by cross-linking interacting RNA species to each other, as has been done for snoRNAs (e.g. U3) and rRNAs (Beltrame and Tollervey 1992).

Experimental identification of new functional RNAs has frequently been achieved and is boosted by genomic sequencing projects. For example, a new type of 200 bp RNA has been revealed in *Mycoplasma pneumoniae* (Dandekar et al. 2000; Göhlmann et al. 2000). Genomic and EST sequencing reveal a huge number of several 100 000 RNAs, as illustrated by the rapid growth of the EMBL database. A large portion of these are only indirectly or incompletely (ESTs) experimentally proven since only the DNA (genome) or fragments (ESTs) have been sequenced. Many catalytic and non-coding RNAs are identified in more direct approaches, such as RNA viridae and their mutants, ribosomal RNAs, and different catalytic RNA sequences including artificially selected ones and autocatalytic introns.

A first experimental task considers the question whether an observed biological effect is caused by a protein or instead by an RNA (without further translation). In the first case, the effects of the newly identified protein would usually be studied in detail - although we point out that there is always a messenger RNA encoding the protein as well as the longer RNA-precursor transcribed in the nucleus of the cell. Thus, strictly speaking, this simplification ignores two further regulatory levels in the cell - the first level operates on the mature messenger RNA and the second on its precursor.

3.2   Typical Challenges

The question of how a biological phenomenon is related to RNA can take the following forms:
- A biological effect may be identified, but there is no indication that it is actually caused by an RNA. Examples of this kind would be the identification of an infectious agent (where it could be bacterium, a DNA or RNA virus or even a prion), or cellular differentiation phenomena. In the first example, the RNA virus may be revealed due to its small size and lack of inhibitory actions by antibiotics. Cellular differentiation phenomena give rise to different phenotypes such as larvae with different shape and impairments. For a number of developmental genes the encoded proteins appeared at first to be the only or dominating cause for the observed phenotype. However, specific localization and other directing signals in the messenger RNAs for these proteins (e.g. *bicoid* mRNA, Ferrandon 1994; *oskar* mRNA, Erdelyi et al. 1995; bicaudal-D, Pokrywka and Stephan 1995) are important in pattern formation in the embryo, in the developmental differentiation and to direct tissue expression of the protein. For new research it is important to be more aware of this, and to design experiments so that locating or other directing and regulatory effects of the mRNA can become apparent and appropriately tested. An independent approach (see Chap. 4) is to look for RNA motifs and regulatory elements as soon as new data or sequences of RNAs from the organism under investigation become available.
- A biological effect is observed, but we have experimental indications (e.g. RNAse sensitivity or sensitivity to micrococcal nuclease of the activity) that a regulatory RNA is already suspected to be a cause of the observed effect. Close observation of potentially involved RNA species and their electrophoretic mobility under different conditions is required. However, often the biological

function of the effect gives a further indication of the identity of the RNA. The next task is to rule out transcriptional regulation and promotor interactions as dominant causes for the phenomenon. This can be achieved, for instance, by Northern analysis, in which constant mRNA concentrations in the presence of varying levels of protein translation indicate the presence of translational regulation.

Well-characterized, regulatory RNAs identified in one system may help to point to similar RNA elements operating in other RNA structures and biological situations in a similar way. This knowledge can subsequently be utilized for systematic computational (see Chap. 4) and for experimental screens, for instance the identification of new iron-responsive-element binding proteins using affinity selections (Melefors and Hentze 1993).

- Deletion and mutation analysis allows dissection of an identified regulatory RNA to examine in more detail how it works. Important functional sites show phylogenetic conservation and appear as conserved regions in a wide range of related RNA molecules, e.g. the Sm binding in snRNAs, conserved catalytic structures in ribozymes or the functional important regions in ribosomal RNAs. Various protocols involving different phages, mutagenic oligonucleotides and thio-nucleotides have been developed for the purpose of site-directed as well as random mutagenesis, though in the meantime constructions by PCR starting with mutated oligonucleotides are easier and faster to obtain in most cases.
Further experimental assays to delineate catalytic and regulatory RNA activities include:
• Debranching enzyme reveals branched intermediates for the action of
trans-splicing RNAs, but also lariat intermediates in *cis*-splicing and self-splicing.
• Detection of lariat intermediates with altered mobility in gel shift assays and usually slower mobility even in denaturing gels.
• Bandshift assays to reveal RNA shifts depending on the concentration of regulatory metabolites.

- A very active way to search experimentally for regulatory RNA structures is to engineer an RNA structure for a desired function. Such experiments concentrate on areas where utilization of RNA molecules is expected to be superior to proteins, for instance different steps of RNA processing. Thus, Bufardecki et al. (1993) undertook a detailed *in vitro* genetic analysis of the structural features of the pre-tRNA required for determination of the 3' splice site in the intron excision reaction.

3.3   Cellular Regulation

Often a regulatory RNA is only one of several possible causes for an observed biological effect. It is challenging to hunt experimentally for a regulatory RNA in this case, but in everyday practice this is the most common situation. The purpose of this book is already well fulfilled if the reader is made more aware that behind the regulatory process he/she is studying, regulatory RNA structures may be involved at critical steps.

As discussed in Chapter 1, regulatory RNA structures should first of all be expected to be important if there is reason to believe that they can outperform proteins in the regulatory step examined. An obvious example is the release of information from the mRNA: examples of regulatory structures before and after (or even in) the coding sequences to determine readout and half-life of the messenger RNA are rapidly accumulating. Further areas of current research are developmental and other complex biological patterns where time (controlling mRNA stability or mRNA translation) and localization signals may be encoded in the RNA, and regulatory networks where RNA may feed into control loops. Finally, if an RNA component forms part of the system under investigation, then it should be examined for both potential regulatory signals and catalytic activity. On the other hand, the probability of revealing new regulatory RNA is low if biological reactions are studied in which proteins are well known to be superior to RNA, for instance rapid catalysis or structural proteins.

A general approach to identify regulatory or catalytic RNA involves bioassays, as exemplified by Cech (1993). During purification (e.g. ultracentrifugation, chromatography and affinity purification), the peak fraction is always identified utilizing the bioassay. If the final, pure fraction contains only RNA, this is a direct proof that the RNA was responsible for the activity monitored. RNAse destroys any RNA activity present in a more or less purified biological activity. The activity result is compared with the result after protease treatment and phenol extraction. This approach led to the identification of RNAse P as an RNA enzyme (Altman et al., 1993). However, the long debate on prion disease as a purely protein-based disease versus a very small viral agent (Coles 1997) shows that a clear distinction between RNA and protein may sometimes be very difficult.

Double-Stranded RNA in Growth Arrest

Regulatory RNAs can sometimes be biochemically identified. Thus, the interferon-induced double-stranded RNA-dependent eIF-2 alpha kinase was shown to have additional functions apart from viral defence (Li and Petryshyn

1991): its activation may be an important signal for growth arrest of fibroblasts and differentiation into adipocytes as its phosphorylation is observed in the absence of viral infection. Total cytoplasmic RNA from 3T3-F442A cells is able to activate the phosphorylation. The activity was purified on oligo(dT) cellulose, thus running with the poly(A)-rich RNA fraction and could be immunoprecipitated with the double-stranded RNA kinase in a specific complex. The activity is double-stranded RNA as it is sensitive to RNAse VI, which is double-strand specific, but not to proteinase K.

Regulatory Structures in Developmental RNAs

Regulatory RNA structures participate in many of the temporal and spatial patterns observed during development (e.g. Stebbins-Boaz and Richter 1997). RNA localization can be demonstrated by in situ hybridization of oligonucleotides complementing the RNA under investigation (Jin and Loyd 1997). Such experiments should be initiated as soon as a clear space distribution of the protein encoded by the mRNA has become apparent, to determine at which level the regulation of this distribution pattern resides.

Monitoring temporal patterns of protein expression and RNA transcription is important to establish translational regulation where the detected protein level changes in spite of constant RNA signals in Northern analysis.

Colocalization by immuno-histochemical or in situ hybridization techniques from cell biology helps to suggest interacting partners of the regulatory RNA. Further tools providing additional evidence for specific interacting partners are selective toxins, for instance against microtubules or microfilaments (Pokrywka and Stephenson 1995). Microscopic techniques give an initial picture of gradient formation and differential development (Kim-Ha et al. 1991). Molecular biology techniques are essential to identify the exact RNA motif, e.g. serial deletions in the regulatory RNA (Kim-Ha et al. 1993), expression of deletion constructs in transgenic animals (Serrano and Cohen 1995) or embryo injection test deletions (Ferrandon et al. 1994).

Regulatory Elements in mRNA

Classical experimental approaches to identify these regulatory mRNA elements quite often used deletional analysis and monitoring translational activity:

Endo and Nadal-Ginard (1987) noted translational arrest in rat skeletal muscle cells. Presence of EGTA, an inhibitor of cell fusion and a potent Ca^{2+} chelator allowed transcription of a battery of muscle-specific mRNAs but not their translation, despite ongoing synthesis of many other proteins, pointing to a specific Ca^{2+}-dependent translational block operating on these mRNAs.

The protein synthesis from the mRNA of the yeast gene CPA1 is repressed if arginine levels are high. The mRNA encodes the small subunit (glutaminase) of the arginine-pathway carbamoyl-phosphate synthetase and the repression is at the translational level. The 5' region of CPA1 mRNA contains a 25-codon upstream open reading frame. Using oligonucleotide-directed mutagenesis and by sequencing of constitutive *cis*-dominant mutations obtained in vivo Werner et al. (1987) and Delbeq et al. (1994) found that the upstream open reading frame plays an essentially negative role in the specific translational repression of CPA1 if arginine levels are high. Three stem-loop structures form in the 5' UTR, two of them are mutually exclusive and the AUG of the leader peptide is involved in base pairing the third (Werner et al. 1987). However, the small upstream reading frame is essential, whereas its exact nucleotide composition is not. Furthermore, this leader sequence can be transplanted and mediates arginine repression also in the context of the GCN4 gene (Delbecq et al. 1994).

Translational control can also be achieved by modulating the rate of translation initiation at upstream AUG codons. This is the case for the yeast gene GCN4 (Müller and Hinnebusch 1986). GCN4 protein is the transcriptional activator for the general control of amino acid biosynthesis. A first pointer to regulatory RNA was the long leader sequence of 600 bp for the mRNA encoding GCN4. Further, the authors noted four very short ORFs, each complete with an AUG and a stop codon separated by two or three codons. Such an arrangement is most unusual, but in general, short open reading frames before the reading frame encoding the protein studied should be examined (regulatory effects of a single short reading frame are shown in Chap. 6). Translational regulation was next easily monitored by growth under different conditions, in this case comparing rich and nutrient-starved media. Under rich conditions the four short upstream reading frames of GCN4 act in *cis* to repress initiation of translation of the downstream GCN4 protein coding sequence, whereas starvation derepresses GCN4. If one wishes to investigate other translational effectors, the correct metabolic conditions in which the maximal regulatory effect is observed have first to be found. All available knowledge about the gene in question should be exploited for this, for instance starvation of a single nutrient (e.g metal ions such as selenium or iron; amino acids such as histidin, tryptophan) or different stress conditions (e.g. heat shock or oxidative stress). Subsequent studies of the GCN4 example established that decreasing the activity of eukaryotic phosphorylation inhibits general translation in yeast but stimulates GCN4 expression as it enables ribosomes to scan past the short upstream reading frames and reinitiate at GCN4 instead. However, there are additional levels of RNA-mediated regulation present in the GCN4 example - and this also should be expected in new studies. GCD10 was subsequently found by genetic screens as an interacting partner of the regulatory RNA. GCD10 acts as translational repressor of GCN4 but turns out to be very probably the RNA-binding subunit of eukaryotic initiation factor 3 (eIF-3; Garcia-Barrio et al. 1995). GCD10 mutations would thus act by decreasing the ability of eIF-3 to induce complex formation of eIF-2/GTP/Met-tRNA (iMet) ternary complexes with small ribosomal subunits in vivo. This explains why mutations in eIF-3 mimic eIF-2 a-phosphorylation in allowing ribosomes to scan through the upstream short reading frames and reinitiate at GCN4. In addition it turns out that sequences 5' of the first upstream open reading frame in GCN4 mRNA are required for efficient translational reinitiation and it remains to be seen whether these additional sequences act on ribosome release or facilitate rebinding of an initiation factor (Grant et al. 1995)

Another example is the *cis*-acting element within the 5' leader of the cytomegalovirus ß-transcript (nts +62- +142). Addition of this signal to an a or immediate early gene construct converted expression of the indicator protein to the ß--class, even though the gene remained under a-transcriptional control. Deletion of a portion of the ß-gene leader sequence reverted expression to the a-class (Geballe et al. 1986).

McGarry and Lindquist (1985) showed by deletion analysis that the 5' UTR leader sequence is involved in the preferential translation of *Drosophila* hsp70 mRNA. During heat shock, translation of normal cellular mRNAs is repressed, while mRNAs encoding the heat-shock proteins are translated at high rates. Specific structural and sequence features of the 5' UTR promote the translation during heat shock of these heat-shock protein-encoding mRNAs.

Simple Northern analysis and determination of protein levels can be a first indication of regulatory RNA operating on translational control: Munro and colleagues (1976) observed that the biosynthesis of ferritin is regulated by iron in the absence of alterations in mRNA levels and thus provided the first experimental evidence for regulatory effects mediated via translational control. The hallmarks of translational control are constant mRNA levels and changing protein concentrations. Stringent controls are necessary in Northern analysis monitoring the mRNA level changes and the corresponding gels monitoring protein levels, for instance an abundant RNA with constant levels such as ribosomal RNA or constant and abundant levels of protein such as actin. The next step is to establish whether the RNA leader sequence mediates the effect (Rouault et al. 1987). The exact sequence requirements of the IRE in the 5' leader sequence of ferritin mRNA were determined using cell culture experiments to monitor iron-dependent regulation of protein synthesis. Constructs were transfected in which different portions of the leader sequence had been deleted, and synthesis of the reporter protein was assayed by CAT activity (Hentze et al. 1987). Similar strategies apply to elements in the 3' UTR but as they often modulate decay of mRNA, the measurement of mRNA levels in exact time courses and under different conditions is particularly important.

A Tripeptide 'Anticodon' Deciphers Stop Codons in mRNA

The two translational release factors of prokaryotes, RF1 and RF2, catalyze the termination of polypeptide synthesis at UAG/UAA and UGA/UAA stop codons, respectively. However, how these polypeptide release factors read both non-identical and identical stop codons remained puzzling. Swaps of each of the conserved domains between RF1 and RF2 in an RF1-RF2 hybrid led Ito et al.

(2000) to the identification of a domain that could switch recognition specificity. A genetic selection among clones encoding random variants of this domain showed that the tripeptides Pro-Ala-Thr and Ser-Pro-Phe determine release-factor specificity in vivo in RF1 and RF2, respectively. An in vitro release study of tripeptide variants indicated that the first and third amino acids independently discriminate the second and third purine bases, respectively. Analysis with stop codons containing base analogues indicated that the C2 amino group of purine may be the primary target of discrimination of G from A. These findings show that the discriminator tripeptide of bacterial release factors is functionally equivalent to that of the anticodon of transfer RNA, irrespective of the difference between protein and RNA.

Screens to Reveal RNA-Protein Interactions

Identification of an interacting protein helps to identify the structural motifs of the RNA involved in this interplay. Rouault et al. (1989) describe a classical method for affinity purification of a regulatory protein that binds specific RNA sequences. RNAs containing the regulatory sequences are transcribed in vitro from oligonucleotide templates, biotinylated and incubated with unfractionated cytosol. Specific RNA-protein complexes are bound in solution to avidin, and the resulting complex is bound to biotin-agarose beads. The cytosolic RNA-binding protein is released from RNA at high salt, and a second round of purification yields an essentially homogenous protein, e.g. the IRE-BP. Variations of this protocol can be applied to various other cytoplasmic and non-cytoplasmic RNA structures.

Methods for the identification of RNA binding proteins include cross-linking of proteins to RNA by chemicals or ultraviolet (UV) light followed by chromatography or density gradient centrifugation. A simplified method (Pelle and Murphy 1993) first induces cross-links between RNA and the proteins which are in close contact to the RNA by UV light irradiation. The extracts from irradiated cells are boiled in the presence of sodium dodecyl sulfate. This disrupts non specific RNA-protein interactions. This is followed by sodium-dodecyl-sulfate polyacrylamide gel electrophoresis (SDS-PAGE), Western blotting from the gel to a filter and washing of the filter. Hybridization of this Western blot with an appropriate nucleic acid probe allows detection of bands containing RNA-protein complexes. Immunizing mice with a region of the nitrocellulose membrane containing the RNA-protein complexes allows to obtain antisera against the binding proteins.

The specificity of cross-links obtained by any such method may be tested by the antisense transcript of the RNA investigated. There should be no interaction with the same set of proteins in cross-linking experiments with this negative control. As the purine / pyrimidine ratio is reversed as well as the sequence order, the use of a random RNA with similar length is another recommended negative control.

In addition, powerful interaction screening systems are available. A good standard is the lexA two-hybrid screening system in yeast which is a powerful means of revealing protein-protein interactions. A version which uses a three-hybrid screen method to hunt for RNA interacting with proteins has been developed (SenGupta et al. 1996). New insights are possible applying in vivo imaging. Thus, RNA-binding proteins may forge the first link in a chain between the mRNA and active transport factors that may include motors, intermediary proteins, vesicles or even organelles. Theurkauf and Hazelrigg (1998) have achieved a real-time analysis of GFP-Exu particle movement in living ovaries. Three types of movements are observed, and two of these provide strong evidence for translocation along microtuble tracks in the nurse cells. For further details see Arn and Macdonald (1998).

3.4   Confirmation of RNA Functionality

Viral regulatory RNA structures can be examined by mutation of different RNA regions. In vivo tests give a very direct proof of functionality. In this way a dual role for the putative dimerization initiation site of HIV 1 in genomic RNA packaging and proviral DNA synthesis has been directly demonstrated (Paillart et al. 1996).

For a wide range of organisms techniques for direct in vivo testing of RNA variants exist. Plasmid shuttling systems quickly test and replace different mutants for complementation of the wild-type gene disruptions, e.g. in fission yeast (Dandekar and Tollervey 1992). Furthermore, synthetic lethality genetically examines protein-RNA interactions in pro- and lower eukaryotes. Plasmid monitoring systems (e.g. red-white sectoring assay in yeast) allow easy detection of synthetic lethality (e.g. Dichtl and Tollervey 1997). This is caused by a combination of two mutations, each of which alone would be without a severe phenotype. The isolated mutations would be far more difficult to detect, requiring more demanding functional tests instead of colony plating. In addition, primer extension analysis of RNA processing intermediates can be recommended as a very sensitive method to detect subtle RNA processing changes in mutants or under changing enviromental conditions. To score allelic variation, Winzeler et al. (1998) show a method that can be used to scan and map any two isolates of a species without allele-specific polymerase chain reaction, without creating new strains or constructs and without knowing the specific nature of the variation. A total of 3714 biallelic markers, spaced about every 3.5 kb, were identified by analyzing the patterns obtained when total genomic DNA from two different strains of yeast was hybridized to high-density oligonucleotide arrays. The markers were then used to simultaneously map a multidrug-resistance locus and four other loci with high resolution.

Detailed Probing of an RNA Motif

As in the analysis of details of protein structure, various mutation, deletion, domain swapping and domain transplantation techniques can and should be utilized here.

Applied to our standard example, iron-responsive elements, a number of studies give experimental protocols for these steps, for instance:

- Establishing distance requirements for iron-responsive elements in the 5' UTR with respect to the coding sequence / translation start (Goosen and Hentze 1992).

- Transplanting a newly identified iron-responsive element to a reporter construct and render the reporter mRNA iron level-dependent (Gray et al. 1995; Dandekar et al. 1991).

- Mutational analysis of stem and loop requirements in the iron-responsive element (Dix et al. 1993).

Ryder and Strobel (1999) adapt a high-throughput screening procedure, nucleotide analogue interference mapping (NAIM), to identify functional groups important for proper folding and catalysis exemplified for the hairpin ribozyme. Their results provide biochemical evidence in support of many, but not all, of the non-canonical base pairs observed by NMR in each loop, and identify the 3' OH functional groups most likely to participate in the tertiary interface between loop A and loop B. The method predicts also non-A form sugar pucker geometries.

Modern gene array techniques allow for large-scale examination of mRNA functionality (various protocols and methods are found in Carter et al. 2000). Thus translational control of mRNA and other regulator effects mediated by regulatory motifs are revealed comparing different mRNA fractions: cDNAs are made either from the polysomal fraction of a sucrose gradient (RNAs are translated) and the messenger ribonucleoprotein (mRNP) or monosome (80S) fraction of the gradient. The different cDNA populations are fluorescent labelled and levels of the original specific mRNAs identified by hybridization to DNA microarrays in different experiments. Different environmental conditions on their effect for translation of different mRNA species can be efficiently compared in this way.

RNA Structure Analysis

Without a detailed three-dimensional structure, a complete understanding of RNA function is not possible - on the other hand, the structure alone is generally difficult to interpret and requires the above-mentioned functional studies to give a comprehensive picture. Improved X-ray techniques reveal detailed crystal structures of the RNA compound. Thus, the 1.8 A crystal structure of a statically disordered 17 bp RNA reveals four distinct conformations of the duplex, with an average pairwise backbone rmsd of 2.35 A. The structural differences between the four conformations, which can be attributed to differences in packing environment, highlight the possible influence of crystal packing forces on nucleic acid X-ray structures. Analysis of interhelical packing between symmetry-related molecules reveals an RNA "zipper" that mediates direct phosphate oxygen-2' hydroxyl interactions between close-packed phosphate-sugar backbones. This may be a general mode for RNA tertiary interaction that does not depend on metal ions or primary sequence (Shah and Brunger 1999). There are rapidly evolving multi-dimensional NMR techniques on RNA solution structures (Laing and Hall 1996). Biophysical and computational methods can be a powerful addition to the structure information gained by NMR. Williams and Hall (2000) approached the difference between prokaryotic tetraloops that are closed by C-G base pairs and eukaryotic tetraloops that are often closed by G-C base pairs. Biophysical properties of the C[UUCG]C and G[UUCG]C were compared using experimental and computational methods. Applying thermal denaturation experiments, NMR data and stochastic dynamics simulations they showed that substitution of a G-C for a C-G closing base pair increases the intrinsic flexibility of the UUCG loop.

Crystallization of ribozymes and small RNA motifs is a difficult task but facilitated by techniques such as the sparse matrix approach (Doudna et al. 1993).

A new method with a large potential is two-dimensional cryo-electron microcopy (reviewed in Dandekar and Argos 1998). New advances in electron microscopy are now possible since digitized electron microscopic images taken from many different viewpoints can be more efficiently processed by computer. New preparation techniques and methods have also evolved such as glucose-embedding and, more recently, ice-embedding and cryomicroscopy. With the aid of experimental automation and computers, it is feasible to obtain a 4 Angstroem map of a helical assembly in which the backbone can be traced (whether it is the backbone of a peptide or a nucleic acid) and a 10-Angstroem map of an asymmetric assembly such as the ribosome in its monomeric state (DeRosier 1997) or, in the context of our research, an RNA motif together with its cognate binding protein, interacting factors and the overall mRNA trace.

Ferre-D'Amare et al. (1998) have developed a crystallization module consisting of a normally intramolecular RNA-RNA interaction that is recruited to make an intermolecular crystal contact. The target RNA molecule is engineered to contain this module at sites that do not affect biochemical activity. The presence of the crystallization module appears to drive crystal growth, in the course of which other, non-designed contacts are made. They have employed the GAAA tetraloop/tetraloop receptor interaction successfully to crystallize numerous group II intron domain 5-domain 6, and hepatitis delta virus (HDV) ribozyme RNA constructs. The use of the module allows facile growth of large crystals, making it practical to screen a large number of crystal forms for favorable diffraction properties. The method has led to group II introns domain crystals that diffract X-radiation to 3.5 A resolution.

Engineering RNA Motifs

The ultimate in understanding a regulatory RNA motif is to be able to engineer it: the motif is so well understood that its properties can be predesigned. For several RNA motifs this has already been achieved. The well-known ribozyme applications for hammerhead and hairpin motifs testify to this. Fortunately for many of the RNA structures well characterized so far, short-range interactions dominate so that a description up to secondary structure and energy stability of different stems and RNA regions is sufficient. On the other hand, rRNA processing and pseudoknots are evidence that for larger RNA structures this is not the case, and long range interactions should also not be underestimated within the mRNA, for instance in the 3' UTR (see Chap. 6).

In addition, RNA motif transplantation experiments (see above) are a simple but effective means to achieve similar command of RNA function and use it for desired tasks in vivo. This also illustrates that there are several other situations apart from direct engineering where task-tailoring of the RNA motif is desired: firstly to show that a suspected in vivo function of an RNA element is indeed coupled to it, secondly, in the opposite scenario, where an RNA-binding protein is known but is partner not yet identified (see affinity screens above) or RNAs related to it may be sought for further experiments and insights.

Another area where RNA structures are sufficiently well characterized to enable direct engineering experiments are ribozymes. Thus, the well-known hammerhead and hairpin motifs (see Chap. 2) enable RNA molecules with catalytic activity to be designed and essential and non-essential features have been determined (Hertel et al. 1996). The work by Szostak's group offers several intriguing examples of the wide range of reactions available to catalysis by ribozymes. For instance, he and his coworkers succeeded in selecting ribozymes able to catalyze amino acid transfer reactions (Lohse and Szostak 1996).

RNA SELEX Experiments

These were initiated by PCR techniques pioneered and established by Larry Gold (reviewed in Türk 1997) and Jack Szostak (Lohse and Szostak 1996). Initial experiments on coupling nucleic acids by linkers go back for a longer time (e.g. Hobbs 1989 and references therein). As an illustrative example, novel RNA substrates for the ribozyme from *B.subtilis* ribonuclease P can be designed and obtained by in vitro selection which comprises the following steps: cleavage of a circular RNA library by the P RNA, isolation of the linear cleavage product and regeneration of circular RNA to allow amplification and multiple cycles of selection. The use of circular RNA ensures that potential substrates can be selected without restricting the location of the cleavage site. Such a selection has been used to isolate RNA motifs that undergo autolytic cleavage with Pb^{2+} (Pan et al. 1994). After eight cycles of selection the circular RNA pool was cleaved as efficient as pre-tRNA Phe substrate, about 10 orders of magnitude better than cleavage of random RNA. The approach worked less well for *E.coli* ribonuclease P which was found to be 10-60-fold less active. Two new artificial substrates were selected (Pan 1995).

Methods for mutation and subsequent SELEX analysis of critical RNA features have been described by Tocchini-Valentini's group. For example different steps of tRNA processing in *Xenopus* tRNA were analyzed in detail, including the intron in this reaction (Baldi et al. 1992; Bufardeci et al. 1993). Work from this group enabled also the construction of a small model substrate for eukaryotic RNAse P (Carrara et al. 1995) and the selection of novel Mg^{2+} dependent self-cleaving ribozymes (Williams et al. 1995). The influence of substrate structure on cleavage by hammerhead ribozymes can be analyzed by this and other similar approaches (Scarabino and Tocchini-Valentini 1996).

Drawbacks of SELEX approaches are not easy to publish and thus are underrepresented in the literature. However, some of the shortcomings should be mentioned here.

In the context of this book one might wish to select and multiply optimal binding features of a regulatory RNA motif involved in binding to the affinity ligand used in the selection scheme. However, besides the selection operating on the criterion of affinity to the ligand used, for instance an RNA-binding protein in an affinity column, there are several other selective mechanisms also operating. Such processes include, for example, the ability of an RNA molecule to act as a template for the PCR polymerase, or the survival of an RNA molecule in the negative screen (for instance, if a washing procedure is used to wash off the column, there will be a selection for general, tight sticking to the support material in the column and bad solubility to the elutant).

Furthermore, good representation of the library can become difficult, not only because the PCR primers set a limitation on the RNA structure finally adopted, but also because in all real applications only a small fraction of all possible nucleotide combinations can be tested (Burke and Gold 1997) and very often the selection scheme is started with invariant stretches of RNA (e.g. high affinity tetraloops for protein binding are preengineered in the RNA population by no mutagenesis in these points). These can sometimes make the search space for the biological optimal RNA sequence inaccessible.

SELEX approaches can be combined with theoretical motif searches (Fig. 3.1). Thus, we investigated sequence requirements downstream of the polyadenylation site by selecting RNA species with high affinity for the CstF protein involved in polyadenylation processing (Beyer et al. 1997).

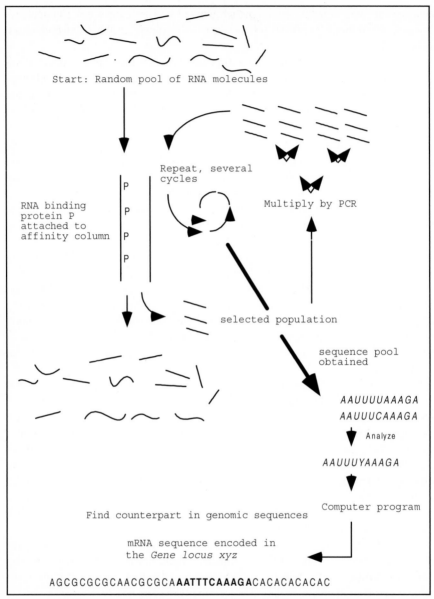

Fig. 3.1. SELEX experiment and retrieval of genomic counterparts. *Top* Typical SELEX scheme enriches by repeated cycles of affinity selection and PCR amplification binding RNAs from a random pool.A cellular protein is used in the column as affinity partner. *Bottom* After sequencing of several RNAs selected common motifs in sequence and structure are defined and translated into a database search program to identify genomic counterparts

In this approach RNA sequences selected in vitro are compared to generate consensus sequences. These consensus sequences can then be checked independently by using them as templates for genomic searches (see Chap. 4). In this way the features found in vitro can be checked for their relevance in vivo. In the example we were able to confirm that several of the motifs identified by in vitro selection are indeed found on genomic mRNAs downstream of polyadenylation sites. The next step would be to use these independently confirmed RNA elements to engineer an artificial polyadenylation site on a test template.

Aleatoric Library

To better understand the intricacies of regulatory RNA, it is very convenient to be able to identify both the RNA and its protein partner. Alan Frankel has developed a method to investigate the interaction of the TAR RNA interaction with TAT. This includes the generation of many random variants (called an aleatoric library) of RNA and RNA-recognizing peptides (Harada et al. 1996). He and his group developed this further to an RNA-binding protein cloning strategy which uses a two-hybrid system-like approach. The RNA to be investigated is cloned into the HIV long terminal repeat in one construct, which thus exhibits transcriptional pausing. This can be relieved only by suitable interaction with the correct RNA binding protein domain attached to the Tat-activator domain protein. *Trans*-activation is monitored using a green fluorescence protein reporter construct. This system is currently being used for the identification of peptides and proteins from different libraries interacting with the Rev-response element. The converse experiment can also be carried out; new RNA sequences can be identified by displaying a binding structure to a given RNA-binding protein domain.

4. Computer-Based and Theoretical Identification of Regulatory RNA

4.1 Introduction

The long list of examples in this book underscores the roles that RNA (regulatory) motifs play in controlling the genetic repertoire of cells and developing organisms. As explained in Chapter 3, identification of an RNA-processing signal, a ribozyme, an RNA localization signal or a translational or mRNA stability control element (examples for each are summarized in Chap. 2, Table 2.2) allows the search for other RNA sequences that bear similar regulatory signals (consensus search). Whilst DNA regulatory elements can often be described by a consensus sequence, RNA signals are frequently composed of combined sequence and structure motifs. A sequence motif is, for example, the hexameric polyadenylation signal AAUAAA. Other RNA motifs are often characterized by complex secondary and tertiary structures that can constitute the major part of the signal, e.g. the clover-leaf pattern typical for the secondary structure of tRNAs or stem-loop structures involved as a second part of the signal in delineating the polyadenylation site in several retroviridae. The search for such motifs and signals is an exercise in understanding RNA structures and sequences as a type of language. As in a human language, both single characters (nucleotides) and higher-order structures (stem-loop structures and more complex tertiary interactions such as pseudoknots) are important. This can also be summarized as interfering contexts of regulatory sequence elements (Trifonov 1996). This chapter first illustrates how searches for new RNA motifs are conducted by computer-aided database screening of genomic sequences and databases for such motifs composed of different sequences and structures. After this we will examine more general (template-free) types of searches.

4.2   Carrying Out an RNA Motif Search

Motif Description

Appropriate formulation of the RNA search target is the first step. For simplicity we will first describe how to search databases in a consensus search: further members of an RNA type of which several examples are already known. Apart from nucleotides that are random or unimportant, several features and nucleotides are essential for the specific function carried out by the RNA within the cell. RNAs which perform a similar task in the cell, for instance in different organisms, cell compartments or enzyme systems, display by evolutionary conservation or by evolutionary convergence similar nucleotides and structures involved in the interactions for this function. To describe and identify such a motif connected with the function and typical for the RNA family, it is first necessary to compile a list of the known family members including important sequence and structure features from literature and direct experimental data.

    Helpful computer tools for these tasks have been developed. Different sequence-retrieval systems retrieve a desired sequence from the database. A classical standard is the GCG software package (Devereux et al. 1984). This is now also available as part of the HUSAR computer package (http://genius.embnet.dkfz-heidelberg.de). This offers a convenient WEB surface and via telnet personal sequence analysis directories. The basic command FETCH copies within seconds the sequence desired from the databank mounted locally on he system (for instance the EMBL database) to your home directory and in this way you can print it out for yourself and read it. This simple system becomes cumbersome if you want to retrieve a whole family, for instance all small nucleolar RNA sequences known in the database. Besides this, very often you do not know the accession number of the sequence under which it is stored in the database. The sequence-retrieval system SRS (Etzold et al. 1996) allows sequence retrieval of whole RNA families (or enzyme families for that matter) by using keywords from the English language that only have to be contained in the database entries which are successfully retrieved. Furthermore, logical operators and phrases can be formulated between keywords and SRS is available on the world-wide-web. Another WEB site highly recommended is the ENTREZ browser from NCBI (http://www.ncbi.nlm.nih.gov/Entrez). This offers a wealth of information including rapid retrieval of literature, sequence information and information on genome sequencing projects.

Common Features

The list of known examples has now to be examined to identify the specific features of the RNA family. In the tables of Chapter 2 we give a number of RNA types or families, including examples of specific family members. Features which define family-specific RNA elements or motifs typical for the RNA family investigated include the primary sequence, the secondary structure and motif position within the RNA (for instance, position in the untranslated region, the open reading frame or intron). The stability of secondary structure within the RNA motif is also important.

All these features are related to the function conferred by the RNA motif to all RNA molecules which carry it. This includes specific protein binding, regulation of RNA stability and involvement in ribosomal processing or translation. However, it is equally important to identify features known to be incompatible with the biological function of the RNA family. Differences in the primary sequence may be known which disrupt the function (mutagenesis experiments); similarly, disruption of stems and loop regions may be known to interfere with the features necessary for the RNA function supplied by the RNA motif. This list of negative features allows counterscreening of the list of putative candidate RNA structures.

Defining a Consensus Pattern

Finally, the compilation of examples and features enables a consensus pattern of the RNA family to be derived (Fig. 4.1). Figure 4.1 illustrates two different types of consensus patterns which we have used in different RNA motif searches for further RNA family members. Thus, the consensus pattern for the *trans*-splicing RNA family (Fig. 4.1, top) is composed of an RNA stem loop containing the catalytic double guanosine just 3' of the loop and is followed by two stem loops flanking a single stranded region which is called an Sm site, because common splicing proteins precipitable by Sm antibodies bind to such a single-stranded stretch composed of two purins on 5' and 3' site flanking a stretch of uracils. Each of the stem loops carries further specific features, apparent after careful examination of the available known examples of this regulatory RNA structure and is briefly illustrated for the important catalytic stem loop on the left in Fig. 4.1. The IRE-consensus pattern (Fig. 4.1, bottom) was derived in a similar way and is analyzed later in this chapter. The collected examples should also be kept (as a file on a computer). They constitute a first test set to check whether the search program (see below) is working.

Possible problems and limitations include: are the family members well defined? In other words, is there now a clear idea of the RNA motif sought after? It should, for instance, be possible at this stage to show a figure of the consensus motif sought (see Fig. 4.1). If not, then the search very probably will not produce a useful result. Another point is that non-updated feature tables may impede the retrieval of the known examples. This is a problem becoming increasingly severe because the databases are so large. It is already now a safe assumption that many helpful hints for retrieval interesting RNAs are buried in databases due to incorrect comments and feature tables. However, in several cases, the knowledge of the biologist conducting the search may rectify overlooked examples and features. A further check on the motif description is to test whether common RNAs such as snRNA, tRNA or repetitive sequences match the family consensus. Again, if they do so in spite of their completely unrelated biological function indicating that the motif description was not sufficiently accurate, the search will fail, as too many unrelated sequences will be pulled out from the databases screened.

Deriving the Search Program

After the available information on the RNA structure of this RNA type has been compiled, features and consensus pattern are transformed into a clear list of features. For this list it must be decided whether a certain feature has to be always present (obligatory) or whether variations are allowed.

Next, weights for the presence of each feature are introduced as well as points for meeting each requirement. This task requires sufficient input from experience, experiment and literature to properly place dominant and less important properties. The negative statements (or weights) on structural features are now also refined to clear statements to exclude certain features and reduce the number of false positive candidates. A minimum score should be defined which is required for the identification of each motif and the complete structure.

This exact description and list of rules is subsequently translated into an appropriate program for a database screen. It is important to realize that this final task can easily be delegated to a programmer, provided that you have previously compiled the list of clear criteria. Furthermore, there are several general-purpose search programs available which can be used to translate the clear list into a screening program. For several RNA types and families there exist specific search programs (see below).

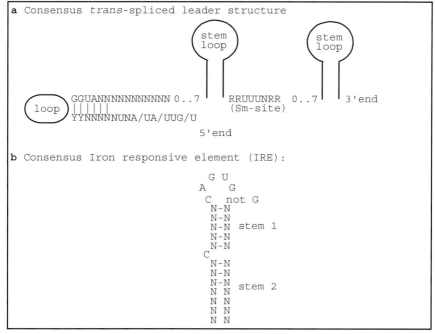

a Consensus *trans*-spliced leader structure

```
                        stem                stem
                        loop                loop


         GGUANNNNNNNNNNNN 0..7    RRUUUNRR 0..7    3'end
loop     ||||||                   (Sm-site)
         YYNNNNNUNA/UA/UUG/U
                             5'end
```

b Consensus Iron responsive element (IRE):

```
                    G U
                 A     G
                 C   not G
                 N-N
                 N-N
                 N-N  stem 1
                 N-N
                 N-N
                 C
                 N-N
                 N-N
                 N-N  stem 2
                 N N
                 N N
                 N N
                 N N
```

Fig. 4.1ab. Consensus patterns in the trans-spliced leader structure and iron responsive element. A search for new regulatory RNAs in genomic sequences can use consensus patterns to identify further candidate structures with the function mediated by the RNA structure underlying the consensus. **a** Simplified consensus structure (detailed consensus in Dandekar and Sibbald 1990). *0..7* indicates none till seven intervening nucleotides, *N* any nucleotide, *R* purines. The leader gets trans-spliced at the guanosine dublett. Compare the consensus pattern with the concrete examples in Fig. 2.11. **b** Simplified consensus iron-responsive element. Compare the more abstract general pattern with examples shown in Figs. 2.5 and 2.7 not G (a negative feature) indicates that any nucleotide from A,C and U is allowed at this position. It is not obligatory for all bases in the second stem to pair; the fifth base pair in stem one is also non-obligatory. The C between the two stems has to be bulged and must not basepair. For effective searching, the typical energy of the two helices is a further filter to select good candidate structures. (Dandekar et al. 1991) Further refinement of this template structure is possible (see text)

Program Tests

Regardless of whether you wrote the search program yourself, had it written by a programmer, constructed a search strategy using a general search program or used one of the custom-written programs, you should now test the program before going ahead and fishing for new RNA structures.

First you should check whether the set of known RNA structures used to construct the consensus can be retrieved from the database. This should be possible (at least for most of them). If you fail here, this indicates that some important feature contained in this type of RNA structure has been overlooked in your feature list.

If you are satisfied with the performance on the test set, you should now try to hunt for already identified RNA examples in the database that were not used in your list to derive the consensus pattern (false negatives). With the increase in database sizes, it should not be too difficult to find such examples upon a full database search.

A particular problem that may arise is that RNA motifs may be overlooked if they are divided between two sequence entries in the database. If this happened to one of the known RNA examples you hoped to retrieve as a positive control, it will never be identified by your program (unless you introduce a special condition to handle this case). Missing such RNA examples is thus no indication that your feature list is incorrect, and can be ignored. In contrast, if the test set is not recognized or a high number of false negatives is present, then this indicates that important features of the RNA consensus pattern have been overlooked or misrepresented. The feature list and, subsequently, the program must first be corrected before proceeding further. Otherwise you are now ready to go ahead and screen databases as desired (and available) with your program.

Program Output

The program output must be carefully reexamined in the light of the biological information available for each potential candidate RNA. It should be borne in mind that the RNA structure, its sequence features or the energy of the stem loop are all only circumstantial evidence for the suspected biological function. Without a suitable biological framework, the candidate RNA structure obtained still has only a low probability to carry this function. On the other hand, the probability increases when the RNA motif search is otherwise very strong in identifying all known examples and has very few false positives (see below). What is mainly sought here is additional evidence from literature, the described function of the RNA and experimental evidence that the RNA structure candidate could perform the suspected function.

Another important point is readjustment of the weights to modify the output quantity. In the first search trials generally the program features demanded are either too strong, so that no new RNA candidates are revealed, or the criteria may be too low, so that a far too large list of potential candidates is compiled. This is modified by changing minimum scores attached to features such as sequence nucleotides required and stem loop stability.

In any case, the feature list should be optimized to reduce the number of overlooked structures (false negatives) already known to carry the function sought in the candidates.

After these refinement steps follows further evaluation of the output obtained by the final implementation of the program and optimized feature list. All available biological knowledge and criteria should now be used as well as the intelligent eye of the observer. It should also be noted whether the candidate motif is conserved between species, for instance the mRNA encoding the same enzyme in different species carries the RNA motif screened for - further independent evidence that the RNA structure found does indeed carry a function.

Promising Candidate Structures and Known Examples

Only the best and most plausible candidates should be analyzed further. Using sequence alignment and structural alignment, they are compared to known members of the family to assess conserved function. Several sequence-alignment programs are available to this end; very useful are ClustalW (Higgins et al. 1996) and the graphical interface ClustalX (Jeanmougin et al. 1998). Caution must be exercised when comparing complete RNA sequences: if long RNA sequences or very different lengths of RNA sequences are compared, the Clustal algorithm does not perform well. In such a case, it is better first to align the sequences moderately accurately using Pileup (GCG suite, Devereux et al. 1984) and then refine this output using Clustal.

To analyze the RNA structure, further programs may be used to check RNA stability and predicted RNA structure. For instance, a predicted or preliminary identified stem-loop structure in an RNA candidate can be checked or further supported by folding the optimal RNA structure after Stiegler and Zuker (using the program FOLD from the GCG suite) or analyzed further in depth for alternative folding possibilities (using the program MFOLD). Different printout programs (e.g. Searl 1993) show the RNA structure in detail. Alternating repeats and other features of RNA structure may be identified using dotplots (Devereux et al. 1984), studying nucleotide entropy alerts for further hidden structural signals (Huynen et al. 1997).

Experimental and Further Tests of the Best Candidate RNA Structures

The candidate RNAs which have passed the last analysis step are now directly tested by biochemical assays. These should first be simple screening assays depending on the RNA type investigated. Thus, if the RNA structure investigated is bound by a protein, this would include band shifts, affinity selection schemes or simple cross-linking experiments. For less clear candidates, competition experiments in band shifts with control RNAs and control-binding proteins are helpful. Catalytic RNA motifs should first be screened for by revealing the RNA intermediate, e.g. as a lariat structure or by using debranching enzyme. Also the folding energy of the RNA is compared to family members and the experimental data obtained (for instance, whether the new RNA structure is as stable in band shifts as are the known controls).

For theoretical analysis, as well as the folding algorithms described above, a sophisticated calculation of the potential three-dimensional structure for smaller RNAs using standard conformations is available (Gautheret and Cedergren 1993); unfortunately, progress in the area of RNA ab initio modelling otherwise is slow. The presence of functional features, such as protein recognition sites (Wittop-Koning and Schümperli 1994), splicing signals (Rio 1993) and polyadenylation sites (Manly and Proudfoot 1994) can be checked.

4.3   Example Searches

New iron-responsive elements (IREs) are a suitable example of such a motif search (Gray et al. 1996; Dandekar et al. 1998). This regulatory RNA element (Fig. 4.1) contains a specific binding site for the iron-regulatory protein (IRP, which was formerly called IRE-BP, IRF, FRP or p90, and of which in the meantime two different versions, IRP1 and IRP2, have been identified). Reduced cellular iron levels increase binding of IRP to iron-responsive elements in mRNA. Binding switches off translation of the ferritin mRNA, but stabilizes the transferrin receptor transcript. As described above, comparison of different RNA structures already known to display the regulatory function examined should lead to a compilation of obligatory and necessary parts of the RNA structure as well as compatible and incompatible features. Figure 4.1 illustrates this (1) for the *trans*-splicing RNAs (top, already discussed) and (2) the iron-responsive elements (bottom). A bulged C-residue in the RNA structure of the iron-responsive element is always present (obligatory feature). The base pairing of the first nucleotides in the "top helix" directly above this bulged C is often observed, but not necessarily required for the iron-responsive element to function (optional feature). Next, relative weights for the presence of each feature have to be introduced. Their respective weight determines greatly the quantity of the output and the specificity of the search. It must be carefully decided which features are judged and weighted to be more important than others. For instance, the base pairs in the bottom helix of the iron-responsive element were scored differently to reflect different stabilities of base pairing (Fig. 4.1, bottom). Negative statements are important to rule out certain features (such as a G-residue in the 6th position of the IRE loop). Provided they are correctly selected, this greatly reduces the number of false positives. The scores (with appropriate weight) for each feature are summarized. A minimum threshold defines when an RNA structure is considered to match the RNA motif and should be defined in advance. Energies for helical structures may also be calculated and scored. For the iron-responsive element a minimum score of 8 (including positive and penalty scores) for the bottom helix (Fig. 4.1) was an obligatory feature to select only stable helices as candidates.

A combined motif search for tRNAs was conducted by Fichant and Burks (1991). Their exhaustive list of obligatory, compatible and non-compatible features included
- the T-Ψ-C signal (2 of 4 invariant bases);
- ability to form the T-Ψ-C arm (loop of 7 bases and stem of 4-5 base pairs);
- similar motifs for D-signal and D-arm, aminoacyl arm and anticodon arm.
Again, a careful definition of a general score was included, which in this case had

to be optimal to minimize both false positive RNA structures (which were, in fact, no tRNA) or false negatives, rejected tRNAs with features far away from the general consensus; 97.5% of the 744 known tRNAs were correctly identified; 42 new putative tRNA sequences were found after detailed analysis of the output. Another screen utilized a custom-written program to reveal and examine the complete set of all yeast tRNAs (Percudani et al. 1997).

The implemented description must always be tested for its search performance regarding positive and negative controls. All mammalian ferritin and transferrin receptor iron-responsive elements in the test set had to be identified in the iron-responsive element search trials. In the automatic identification of group I intron cores (Lisacek et al. 1994), a set of 93 intron sequences provided the initial test set and had to be retrieved.

4.4   Motif Specific Searches for RNA Elements

These, as well as the general RNA search programs, are an incentive to look for new regulatory RNA elements. Many of them are available from the respective authors, some are also directly available via FTP. Combined motif searches consider both sequence and structural motifs to identify members of an RNA family. Other related algorithms and approaches for each application are cited in the original publications.

Searching for tRNA Genes in Genomic Sequences

Fichant and Burks (1991) predicted 42 previously unidentified tRNAs. The program (compare with above) is a good illustration of a carefully defined combined motif search. It had a very low false positive rate (0.003%) and was developed for genome analysis.

An approach applying covariance models performs even better, in particular if diverged tRNAs have to be screened (Eddy and Durbin 1994). The covariance model constructs a tree connected by transition probabilities to generate the known sequences of a family and searches with this covariance model for new unrecognized members. Though such an approach may be suitable for a general search method, its limitation is currently slow speed, as only 10-20 base pairs per second can be searched.

Percudani et al. (1997) successfully retrieved all yeast tRNAs using not covariance methods but again a careful list of required and non-required features.

Prediction of Transcription Terminators

Ermolaeva et al. (2000) analyzed rho-independent transcription terminators in bacterial genomes. They developed an algorithm that further evaluates the accuracy of the predictions. Twelve bacterial genomes were searched using a common mRNA motif: a hairpin structure followed by a short uracil-rich region. For each terminator an energy-scoring function that reflects hairpin stability, and a tail-scoring function based on the number of U nucleotides and their proximity to the stem were computed. Thereafter, a confidence value reflecting an empirical estimate of the probability that the sequence is a true terminator was calculated. In 4 of the 12 genomes they found plenty of rho-independent terminators and, using *E.coli,* they calculated the sensitivity of their method ranging from 89 to 98%, with corresponding false positives of 2 to 18%.

Identification of Catalytic Introns

The different structure and sequence signals contained in catalytic group I introns were translated into a combined motif search. Nearly all (132 out of 143) previously known group I introns were correctly identified by the algorithm. However, no new, previously unidentified group I introns were revealed in this publication (Lisacek et al. 1994), but with a very low false positive rate of 1 per 1 million base pairs the program can be useful for genome analysis of new sequences.

Candidates for trans-Splicing RNAs

We carried out a combined motif search considering primary sequence, secondary structure and energy as well as postive and negative obligatory and non-obligatory features to screen for possible candidate RNA structures (Dandekar and Sibbald 1990). The consensus features are depicted in Fig. 4.1. We revealed previously unidentified RNAs capable of *trans*-splicing or with the potential for a similar catalytic activity. The search successfully identified all genera known to *trans*-splice and most of the known *trans*-splicing sites. Negative controls were rRNA, snRNA and tRNA, which were only misrecognized in a few instances. Subsequent experiments verified several candidate structures. Interestingly, *trans*-splicing reactions could experimentally be carried out in HeLa cells (Bruzik and Maniatis 1992) and several vertebrate RNAs are suggested to be involved in *trans*-splicing reactions (Chap. 6).

Iron-Responsive RNA Elements

A combined motif search includes different features (Fig. 4.1) of sequence, secondary structure and energy to identify new iron-responsive elements in the

5' UTR of different mRNAs (Dandekar et al. 1991, 1998; Gray et al. 1996). The consensus used for the searches and the feature list is being further refined as additional experimental data clarify the features of the IRE in regulation. Combination with accurate experimental testing is critical. Besides different bandshift and competition experiments, this also includes testing the function of the IRE motif in other mRNA contexts, for instance to regulate translation from a reporter gene construct in an iron-dependent manner.

U-Turns in Ribosomal RNA

The U-turn RNA motif is characterized by a sharp reversal of the RNA backbone following a single-stranded uridine base. Gutell et al. (2000) analyzed U-turns in ribosomal RNA (rRNA) with comparative sequence analysis in 16S and 23S rRNA sequences. They found 34 U-turn candidates occurring in hairpin loops and 24 U-turn candidates in multistem loops. In 13 cases, the bases on the 3' side of the turn, or the 5' immediate 5' side, are involved in tertiary covariations, making these sites strong candidates for tertiary interactions.

Methylation Guide Small Nucleolar RNAs (snoRNAs)

Ribose methylation is a prevalent type of nucleotide modification in rRNA. Eukaryotic rRNAs display a complex pattern of ribose methylations amounting to about 100 in vertebrates. Each is guided by a cognate small RNA. Gaspin et al. (2000) searched archaeal genomes for potential homologues of eukaryotic methylation guide snoRNAs, combining searches for structures motifs with homology searches. They identified a familiy of 46 small RNAs conserved in the genome of hyperthermophile *Pyrococcus* species. They suggest that such small RNAs may play in addition an important role in rRNA folding. Lowe and Eddy (1999) applied probablistic modelling methods akin to those used in speech recognition and computational linguistics to screen the yeast genome for snoRNAs; 22 further methylation guide snoRNAs, labelled snR50 to snR71 were identified; 51 of 55 ribose methylated sites in yeast ribosomal RNA can now be assigned to 41 different guide snoRNAs.

Identification of RNA Pseudoknots

These RNA structures are three-dimensional structural motifs and particularly challenging. A motif search by Chen et al. (1992) utilized a Monte Carlo procedure to extensively test the stability of putative pseudo knots. Four out of five predicted pseudoknots from TMV RNA could be confirmed by experiment. The program also succesfully identified pseudoknot structures from other viral sequences.

Non-nearest neighbour effects on the thermodynamics of unfolding of a model mRNA pseudoknot were determined in phage T4. In this study, the equilibrium unfolding pathway of a 35-nucleotide RNA fragment corresponding to the wild-type and sequence variants of the T4 gene 32 mRNA has been determined through analysis of dual-wavelength, equilibrium thermal melting profiles via application of a van't Hoff model based on multiple sequential, two-state transitions. Further details are found in Theimer et al. (1998).

Searching for the Canonical aauaaa Element

Graber et al. (1999) investigated mRNA 3'-end-processing signals in six eukaryotic species through analysis of 3'-expressed sequence tags. The conservation of the canonical AAUAAA element differs widely among the six species and is especially weak in plants and yeast. The complex pattern of the use AAUAAA and single-base variants of AAUAAA indicates that effective tools to identify 3'-end processing signals requires more than consensus sequence identification.

Common Secondary Structure Motifs of Homologous RNAs

Han and Kim (1993) predicted common secondary structure motifs. A phylogenetic alignment of several sequences provides the start (compare with the program coresearch, Wolferstetter et al. 1996; below). From this, a covariation matrix is formed. Non-overlapping, conserved stable helices eliminate alternative, less stable helices to calculate the common most stable fold. The performance of the algorithm was evaluated with tRNA, 5S rRNA and 16S rRNA. Structures that may possibly be involved as packaging sequences in HIV-1 were identified by this search.

Bouthinon and Soldano (1999) invented a new method for predicting the secondary structure of unaligned RNA sequences. In a first step, a best secondary structure is created, applying energy minimization and represented as a structural pattern. The next step involves searching for repeated structural patterns and thereafter the plausibility of each repeated structural pattern is computed by checking if it occurs more frequently than in random RNA sequences. With this method the authors found putative consensus structures for tRNA, fragments of 16 S RNA and 10 Sa RNA.

5' Splice Sites in Pre-mRNA

This is a sequence motif search. Specific common nucleotides around the 5' splice site were classified into 33 subclasses by Kudo et al. (1992). The scheme was used to predict 5' splice sites in new sequences with 90% certainty. Better prediction of 5' splice sites and introns is an intriguing focus of research. Large-scale sequencing projects depend on a better identification of reading frames (see below).

5' UTR in Poliovirus

Currey and Shapiro (1997) use an extensive parallel computer architecture to identify secondary structure in the poliovirus 5' non-coding region by a genetic algorithm. The main benefit of this approach is that the genetic algorithm mimics in an abstract way evolutionary constraints and thus predicts the observed secondary structure without need of phylogenetic comparative sequence analysis required if dynamic programming algorithms such as those by Zuker (1989), Zuker et al. (1999) are used. The secondary structure of the region containing the determinants of neurovirulence was better predicted. This result is interesting, but has to be interpreted with caution, as it is the first example of its kind. However, it may be that the need for phylogenetic comparisons and beforehand knowledge to correctly predict secondary structure of larger RNA molcules may be reduced by such approaches.

Analysis of base pairing between 16S RNA and 5' UTR for translation intiation
Osada et al. (1999) generated an algorithm for analyzing the free-energy values of the base pairing between the 3' end of 16S RNA and 5' UTR of mRNA, in order to analyze the base-pairing potentials in various prokaryotes. The algorithm can be used to comparatively examine the mechanism of translation initiation. The C program used is available upon request to the authors.

Non-Consensus Searches

Searching with a consensus template, new RNAs with regulatory elements can be identified in genomic screens. If an RNA element is not yet identified, it may be analyzed by combining results from SELEX (selective enrichment of ligands by exponential amplification) and a search of databases from RNA or genomic sequences. We delinated a novel instability element in the 3' UTR of the Estrogen receptor mRNA in this way (Dandekar et al. 1998). This is only one example for this quite general class of RNA element searches (see some more examples below).

Prediction of mRNA Translatability

Kochetov et al. (1999) determined sets of 5' UTR characteristics, significantly different between mRNAs encoding abundant and scarce polypeptides, for mammals, dicot plants and monocot plants, and collected them in the LEADER_RNA database. These computer tools for predicting mRNA translatability are available at: http://wwwmgs.bionet.nsc.ru/programs/acts2/

4.5 General-Purpose Search Programs

General-purpose search programs allow interactive translation of the RNA family description into a program. However, they are neither as tailored nor as limited as a fine-tuned specific search program. In particular, if structural motifs have to be scored and evaluated, a specific search program is usually better suited. The following general search programs are sorted according to their level of complexity. Each of the listed programs can be obtained from the original authors or is directly available via FTP. Efficient use of any of these requires the steps of refinement and testing discussed above. Thus, in a certain sense, they present high-level programming languages tailored to screen for RNA motifs which still have to be properly programmed and compiled to yield an efficient search program.

Findpattern

Findpattern is a simple but versatile program, it forms part of the GCG program suite (Devereux et al. 1994). It runs on the VMS operating system and the VAX, but is also available and running on systems with the UNIX operating system. As it is is easy to use, it is highly recommended also for beginners. Sequence motifs, including wobble positions and mismatches, can be extracted from databases. Less obvious and of great value to the user, is the fact that suprisingly complex patterns can be sought, for instance:

GRYGC(G,AG){1,5}AW~(C,T)T(N){14,20}TGT,

which represents G, followed by a purine and a pyrimidine, followed by GC, followed by either G or AG, repeated 1,2,3,4 or 5 times, followed by A, followed by A or T (W represents a wobble codon), followed by a nucleotide which is *not* C or T (~ means *not*), followed by T, followed by a stretch of 14 to 20 nucleotides, followed by TGT. The list of abbreviations meaning several of the four possible nucleotides (or amino acids, for that matter) at one position can be easily found in the GCG program by typing the command Appendix iii or Appendix iv.

The output of FINDPATTERN identifies the wobble positions for each hit, which is also very useful when looking for the most perfect match and if you have some knowledge (from your experiments!) of where the more important positions of the motif are positioned. A limitation is that only sequence motifs can be searched for, thus secondary structure and stem loops are not easily identified. Furthermore, no threshold scores can be defined; anything matching is a hit. For many simple applications this is, however, sufficient, as fine tuning the number of mismatches using the option /mismatch= decides on the quantity of the output and thus stringent and less stringent searches can be conducted in a simple way.

Patsearch

PatSearch is a program that finds functional elements in nucleotide and protein sequences and assesses their statistical significance. Pesole et al. (2000) generated this software, which can also search for complex nucleotide patterns including potential secondary structure elements, also allowing for mismatch / mispairings below a user-fixed threshold and using a Markov chain simulation to assess statistical significance of their occurrence. Neat is also the suite of different databases offered to search with the pattern. A web interface is located at: http://bigarea.area.ba.cnr.it:8000/EmbIT/Patsearch.html

Overseer

RNA motifs composed of strings with and without mismatches are searched by this program. Repeats, palindromes, single positions and interacting positions compose the searchable motifs (Sibbald et al. 1992). Palindromes may also yield RNA helices. In this way, the program identifies secondary structures. Loop length as well as stem size can be specified. More complex RNA structures are interactively assembled. One building block after another is defined in dialogues. A model dialogue is described in Sibbald et al. (1992). If you intend to use this or the following general-purpose search programs for motif searches, the different steps of compiling a list of features, checking for positive and negative controls, output size etc. as outlined above are essential.

An interesting feature of this program is that sequence titles may be searched at the same time as the sequence title is also treated and searched as a string (albeit with more rich characters than an RNA molecule). Limitations of overseer are that specific weights or a scoring scheme for the building blocks cannot be defined. Furthermore, mismatches in the stems are allowed, whereas bulges are not. As RNA structures such as TAR (transactivation response element) from HIV or the iron-responsive elements depend in their function critically on the presence of certain bulges, this limitation can be serious. Also the energy of stems or other structures is not considered, and building blocks are obligatory. Complex building blocks like pseudoknots can thus not be incorporated. However, despite its limitations, the dialogue language used to build up structure search templates makes this program quite useful for RNA secondary structure searches, even in large databases, provided the RNA structure is not too complex.

The AMIGOS Program

A systematic method has been developed for classifying and analyzing the variety of conformations adopted by nucleic acids. The AMIGOS program can calculate pseudotorsion angles (eta and theta), together with a complete library of conventional torsion angles, for any RNA structure or all-atom model. Having computed eta and theta for each position on an RNA molecule, they can be represented on a two-dimensional plot representing the conformational properties of an entire RNA molecule. This facilitates rapid analysis of structural features (Duarte and Pyle 1998).

The ANREP Search Program

ANREP (Mehldau and Meyers 1993) identifies patterns composed of spacers and approximate matches to motifs. These are recursively defined and finally composed of "atomic" symbols. A drawback is that the user has first to learn the user language A. However, complex sequence motifs can then be composed with motifs separated by different spacer lengths. The major advantage of this program is that an elaborate scoring scheme, thresholds and even scoring matrices of allowed nucleotides in certain positions for each motif can be included, permitting also insertions and deletions with respective weighting.

This program searches DNA and protein sequences with equal ease. An unfortunate general limitation is that this program does not identify structural motifs, although RNA structure is connected to many of the more interesting RNA motifs.

Staden's Program Suite

The philosophy behind the original program (Staden 1989) is to create a dictionary of "nucleotide words", e.g. occurrence of TATATATA, and to compare also closely related "fuzzy" sequences (e.g. three mismatches are called by Staden a fuzziness of 3). This program can be used to find common nucleotide motifs hidden in an RNA family and thus help to reveal a consensus motif. The limitation is that common structures may be missed. However, this old program can already complement the first three types of searches. However, this is only part of a whole suite of motif search programs which are constantly being advanced and further developed by Staden (1998). Though not always easy to use, these programs can be surprisingly powerful and cover restriction site searching, translation, pattern searching, comparison, gene finding and secondary structure determination. These programs require computers running the UNIX operating system. Detailed information about the package is available on the web site http://www.mrc-lmb.cam.ac.uk/pubseq/.

Cleanup

This program (Grillo et al. 1996) enables the removal of redundancies from nucleotide sequence databases. An algorithm based on an approximate string-matching procedure removes sequences considered as redundant if they show a degree of similarity between an overlapping longer sequence already in the database greater than a threshold fixed by the user. We are in an age of rapidly accumulating databases, where considerable overlap is easily produced, for instance in EST-sequencing efforts. Such or related tools become more and more important (as much as improved capabilities to carry out unbiased statistics).

Coresearch

Wolfertstetter et al. (1996) offer a general-purpose algorithm to identify potential functional elements directly from nucleotide sequence data (thus for both DNA or RNA sequences). A set of at least seven not closely related sequences with a common biological function is required; however, it is only necessary to know that the function is correlated with one or more unknown sequence elements present in most but not necessarily all of the sequences. The program defines N-tuples. A tuple in the context of this program is a group of nucleotides occurring in the sequences. The N denotes how many nucleotides are involved in the group. N-tuples which occur at least in a minimum percentage of the sequences with no or one mismatch (at any position in the tuple) are calculated. Selection is carried out by maximization of the information content first for the N-tuple, then for a region containing the tuple and finally for the complete binding site. Further matches are found in an additional selection step. The algorithm is able to delineate short-sequence core motifs in sets of unaligned sequences of about 500 nucleotides using no additional information. Though more complex RNA motifs can be missed by this approach, and it is particularly limited if only short stretches of template RNA molecules are known, the program is a promising first start to identify regulatory motifs without any additional work except feeding in sequences which carry out similar biological function.

Further Search Tools

Other computer tools which may prove helpful include:
Covariance methods (see above) can also be applied to other RNA motifs. Advances in secondary structure prediction include multiple alignment using simulated anealing (Kim et al. 1996). Extensive comparative sequence analysis protocols have been described by Stormo's group (Gutell et al. 1992). A good introduction to sequence analysis tools can be found in Doolittle (1990). A new and rapidly increasing source of information is the World-Wide-Web. Several of the researchers mentioned are also interested in making their tools available via this new communication server, for instance the sequence-retrieval system is available on the net. Nearly all of the major databases are now accessible by web interfaces and this can further enhance the speed of identifying related RNAs.
Very useful is the Vienna RNA package (original available at http://www.tbi.univie.ac.at/~ivo/RNA/). Besides this, there are also several sites with various RNA databanks, for instance at Baylor College of Medicine in Texas, http://mbcr.bcm.tmc.edu/smallRNA/smallrna.html, where different RNA families are compiled by Ram Reddy and his group, or at the University of Utah where a

database ODNBase (Giddings et al. 2000) on successful antisense oligonucleotides for mRNA targeting is available (http://antisense.genetics.utah.edu/). Table 4.1 summarizes some further links.

Table 4.1 Some program and database WEB links

Databases	
EMBL database	ftp://ftp.ebi.ac.uk/pub/databases/embl/release/
Genomes of different species	ftp://ftp.ncbi.nlm.nih.gov/genbank/genomes/
RNA families	http://mbcr.bcm.tmc.edu/smallRNA/
OBDNbase	http://antisense.genetics.utah.edu/
RNA-folding programs	
Mfold	http://bioinfo.math.rpi.edu/~zukerm/
Vienna RNA package	http://www.tbi.univie.ac.at/~ivo/RNA/
Different programs	
mRNA translatability prediction	http://wwwmgs.bionet.nsc.ru/mgs/programs/acts2/
GCG Wisconsin package	http://www.gcg.com/products/wis-package.html
Patsearch	http://bigarea.area.ba.cnr.it:8000/EmbIT/Patsearch.html
Overseer among many others	ftp://ftp.ebi.ac.uk/pub/software/unix
AMIGOS	http://cpmcnet.columbia.edu/dept/gsas/biochem/labs/pyle/research/computation/structure/
ANREP	http://www.cs.arizona.edu/people/gene/
Cleanup	http://bigarea.area.ba.cnr.it:8000/EmbIT/coda_clean.html
Coresearch	http://www.gsf.de/biodv/coresearch.html
Staden's program suite	http://www.mrc-lmb.cam.ac.uk/pubseq/
Alignment programs	
BLAST familiy	http://www.ncbi.nlm.nih.gov/BLAST/
Clustalx / Clustalw	http://www-igbmc.u-strasbg.fr/BioInfo/ClustalX/Top.html
Useful links	
Many RNA links	http://www.imb-jena.de/RNA.html
Many RNA links	http://www-lbit.iro.umontreal.ca/RNA_Links/RNA.shtml

4.6   Evaluation of the Search Result

The program output is now analyzed exploiting all the biological information available for any of the candidate RNA structures. Good search programs allow for fine-tuning. If this feature is available, the relative weights attributed to each feature can be readjusted to improve the output; in addition, the threshold score also determines the quantity of the output. Only a limited number of candidates can be tested further. Thus, the new candidates should be ranked in order of a putative biological role for the newly identified RNA motifs using additional criteria and biological knowledge. An objective ranking is difficult to achieve, but is not strictly necessary, because this step serves only to reduce the RNA structures for the more time-consuming experimental tests. Thus, for three top candidate RNAs where we successfully identified new iron-responsive elements after database searches, biological evidence indicated the functionality of the RNA structures in the following way according to evidence from laboratory experiments and literature: erythroid 5-aminolevulinate synthase is involved in haemsynthesis and iron utilization. Succinate dehydrogenase and mitochondrial aconitase are both iron sulfur proteins and hence depend on suitable iron levels in the cell. Furthermore, aconitase displayed significant amino acid similarity to the IRE-binding protein IRP.

With a very high threshold score, only the original test set may be (re-)identified yielding no further information. However, a low number of false positives is important if large amounts of sequences have to be screened, e.g. tRNAs in genome sequencing (Fichant and Burks 1991). A low threshold score drowns interesting candidates in a background of hundreds of false positive hits. A good rule to set the threshold score is the maximal score at which still nearly all known members of the RNA family in the databank may be retrieved which were not part of the first test set of sequences. In IRE searches (Dandekar et al. 1991; Gray et al. 1996) the threshold score for the bottom helix was set at 8, scoring G-C pairs with 3, A-U pairs with 2 and G-U base pairs in the bottom helix with 1 point. Higher scores no longer permitted the retrieval of all known IREs (Fig. 4.1, bottom). When the score was lowered by 2, the output more than doubled without yielding more candidates for which biological evidence encourages direct experimental testing. With good positive and negative controls, many new hits point to the existence of a large RNA family. Examples for this are *trans*-splicing RNAs (Dandekar and Sibbald 1990; Nilsen 1995). The feature list may be refined and the screen repeated after shifting weights from less specific to more specific features as they can be identified from known family members which were not part of the test set. A crude description of the *trans*-splicing RNAs was thus adjusted after the first searches to yield the more detailed description shown in Fig. 4.1. If the program is able to identify the majority of the known structures without identifying new candidates near the threshold score, no further family members may be stored in the database.

Further evaluation of the output by eye and available biological criteria provides additional screening filters. For instance, due to their defined roles in the regulation of mRNA translation (Dandekar et al. 1998), IREs were only further considered if found (1) in RNA which is mRNA and (2) within the 5' untranslated region of this mRNA and (3) were not previously identified. Another example of a detailed analysis of an RNA search output by statistic, thermodynamic and biological criteria is described for potential RNA pseudoknots by Chen et al. (1992). A third more quantitative example illustrates searches for polyadenylation motifs (Table 4.2).

Table 4.2. Reducing the number of RNA candidate structures from the database search

Example: Output of a database search for a *polyadenylation site downstream element*

Step	Searching for	Total sequences identified
AAUAAA	polyA sites	Tens of thousands (in most eukaryotic genes)
multipartite downstream regulatory element	100 - 300	(depending on the motif complexity and data base size, see Figures 2.3 to 2.6 for a number of different motif examples)
1st filter RNA ?		(Alternative: non transcribed, DNA)
mRNA ?		Several sequences
2nd filter Position ?		(Alternative: coding region, upstream)
3' UTR ?		Few sequences
3rd filter Known ?		(From literature; positive controls)
New ?		Selected candidates (e.g. 3) for experimental testing

The Best Candidate RNA Structures

The next step is the comparison of newly identified candidate motifs with known RNA family members. Thus, a new selenocysteine mRNA motif should be closely compared with the known examples shown in Fig. 2.4. Alternative secondary structure folds of the candidate RNAs should also be considered, e.g. by using the program mfold (Devereux et al., 1984). Looking at different IREs, the subsequently identified new iron-responsive elements from eALAS, aconitases or *Drosophila* succinate dehydrogenase resemble the genuine iron-responsive element from human ferritin chain-IRE (Fig. 2.7). Phylogenetic conservation of a newly identified RNA structure, e.g. in mRNAs of different species encoding the same gene (e.g. similar enzyme activities), is a further confirmatory piece of evidence. Thus, the rev-response element-like structure identified for human DNA ligase I mRNA is also found in yeast and mouse (Dandekar and Koch 1996). The comparison yielded a similarly clear picture in the case of the tRNA search by Chen et al. (1992), but new pseudoknot candidates were not so easy to compare. Hidden similarities between known members of an RNA family and new candidate RNAs can be better recognized by dot plots (Devereux et al. 1984) and correlation images (Nedde and Ward 1993). An example of a correlation image is the comparison of a newly identified small nuclear RNA with a known small nuclear RNA: a horizontal line indicates that the newly identified RNA is probably related, small bars indicate further regions of similarity (program available, Nedde and Ward 1993). Structural motifs like hidden repeats within the sequence may also be revealed in this way (Milosavljevic and Jurka 1993).

Experimental tests of candidate RNAs include detection of branched intermediates in *trans*-splicing RNAs. These are analyzed in vitro using debranching enzyme or by their anomalous migration in two dimensional gels (Nilsen 1995). Band-shift experiments can be used to assess binding of a protein to a new putative RNA family member. The newly identified aconitase and *Drosophila* succinate dehydrogenase IREs were experimentally confirmed using this and other methods (Gray et al. 1996). In contrast, the toll IRE motif did not bind functionally stable and also calculated to be less stable. Also four out of five pseudoknots found by the RNA search effort of Chen et al. (1992) with the 3'-terminal non-coding region of TMV RNA were experimentally confirmed, together with several other new examples.

After a first experimental confirmation more elaborate techniques can investigate the exact function of the RNA element. For instance, transplantation experiments can be performed, in which the RNA element is placed upstream of a completely different reporter construct to give more firm proof of the specific function of the RNA structure. In addition, RNA regulation can then be tested under different metabolic conditions (particularly time-demanding in the in vivo situation and the complete organism) and so forth.

Table 4.3 gives a quick checklist on important points during an RNA motif search.

Table 4.3 Some points to check during an RNA motif search

- False negatives
- False positives
- Too many hits
- Is the query sequence retrieved ?
- Test set not recognized ?
- Split entries in the database !
- Biological significance ?
- Suitable experimental tests

Generalized Searches for RNA Motifs

Several biologically important RNA families have already been successfully extended by the identification of new family members from databases. As we have indicated in the above description of protocols and available software, it is now clearly feasible to apply computational approaches as powerful tools for the identification of RNA motifs. On the other hand, a motif search is not a magic tool. As indicated by the necessity for different validation steps described above, even a search with a general purpose or highly tuned search program needs a careful evaluation of its performance, stringency and of the newly found candidates, including more or less elaborate experimental tests.

A further level of complexity is added by the context in which each regulatory motif is contained. In general, on each particular piece of RNA several different codes overlap (several in a particular position within the RNA though not all at the same time) or impose restrictions on how an additional signal specifically carried by this particular RNA can be phrased. This also implies that most of these additional codes have some degeneracy. For RNA (in contrast to DNA in the nucleus, Trifonov 1996) we could mention, for example:

- Classical triplet code (if a reading frame is considered)
- RNA shape code
- Nuclear export code
- Splicing and processing signals (including polyadenylation)
- Translational signals (including initiation and termination and speed of translation and influence on protein folding)
- RNA stability restrictions
- Further undiscovered features.

A clear theoretical concept is hence critical for a computer-based motif search to succeed and concentration on database aspects and programming is helpful. However, in our experience, the combination with experimental data and incorporating all available experimental information is equally important. Possibilities for motif matches or RNA secondary structures and base pairings are generally too high, as the important additional information required to prune them using selected experimental information may be neglected. This is illustrated in searching for the correct base-pairing possibilities of small nucleolar RNAs with different regions of precursor rRNA, where the multitude of potential base pairing potentially assigned by computer can be quite efficiently eliminated in the light of experimental data, for instance for the regions of base pairing between U3 RNA and the yeast ribosomal repeat (Beltrame and Tollervey 1995).

A classification scheme for complex interactions focusing on three-way and four-way junctions in particular is reviewed and explained by Altona (1996).

As more and more instances of regulatory RNAs become known, less stringent searches are becoming possible. In fact, it is also possible to search known sequence information for novel regulatory motifs without a strict definition of motif templates. Thus, broad screens for catalytically active RNA structures are possible. Another area where very flexible criteria and motif search programs are being sought for and developed is the recognition of intron-exon boundaries, of splice sites and the detection of open reading frames. Numerous approaches have been developed for this purpose (see some of the examples mentioned above). In particular, a promising example is the application of Bayesian probabilities and hidden Markov models for refined RNA motif searches allowing for discrete events such as gap or shift. Neural networks present another attractive way of defining an RNA structure by training on the set of RNAs known to possess this structure. They mimic computationally the pattern recognition by neurons in living organisms. Connections between the modelled neurons are not preprogrammed but instead selected for optimal performance in the recognition of a training set where the correct answers are all known. Subsequently, the optimal trained network is used to detect as yet unrevealed patterns. The hidden "layers" of the network (all the internal connections of the net except input and output) may be further analyzed to unravel analytical rules for the set of functional RNA structures analyzed by the neural net. Examples of neural net recognition of RNA motifs include splice site identification (Ogura et al. 1997), short functional regions such as promoter regions (Giuliano et al. 1993) and general identification of reading frames (Solovyev and Lawrence 1993). However, the multitude of approaches illustrates that this and other non-consensus searches show space for further improvement.

Both the theoretical and experimental approaches discussed in the last two chapters can also be modified to look for more general RNA regulatory structures; an example is our non-consensus search for an instability element in 3' UTR regions of mRNA mentioned above (Dandekar et al. 1998). Thus, apart from distinct motifs in the untranslated region of mRNA, one can envision general screens of untranslated RNA regions which give scoring points for the potential to be used as functional RNA structure. Similar, catalytic potential in highly structured RNA may be estimated (Dandekar and Sibbald 1990). Also the experimental techniques allow a wide range of more open or less stringent searches for functional RNA, for instance in different affinity selection schemes, or "open-ended" SELEX schemes in bioreactors where different catalytic RNA populations compete for maximum survival on comparatively loose obligatory structure requirements. Though such work is a contemporary research focus (Ehricht et al. 1997) such experiments had already been started in the late 60s and early 70s (Mills et al. 1967, 1973).

In general, the combination of both approaches, searching experimentally and by computational tools, will be even more powerful.

Another type of non-consensus search to mention is the following strategy: start from one RNA molecule alone, not constructing elaborate templates or consensus sequences but instead look for direct relatives to the sequence by different homology searches. Such a direct "match and hunt" approach has proved to be suprisingly powerful, for instance in further characterization of yeast U3 RNA sequence and promotor region (Dandekar 1991).

The homology of several highly conserved regulatory RNAs is high enough to allow direct identification of relatives by fast motif-searching procedures such as BLAST or BLAST2 (BLAST2 allows for gaps and deletions). Reverse blast searches offer a further control for such approaches, and a means to evaluate the significance of candidate hits found by such a strategy. This involves testing whether the candidate with high homology is able to retrieve the original query sequence again with significance (Bork and Gibson 1996). The identification of snoRNAs provides a case in point: probabilistic modelling methods allowed retrieval of 22 further methylation guide snoRNAs in yeast (see above; Lowe and Eddy 1999) and, more importantly, consensus searches identified 46 small RNAs conserved in Archaea (Gaspin et al. 2000).

Even very general screens which identify all potentially interesting structures in the 5' UTR or 3' UTR regions of mRNA are possible, albeit at the cost of very rapidly accumulating a large set of potential structures which have to be tested.

Template-free searches without a database example can also be achieved by the following three methods:
- Incorporation of different experimental data to derive an RNA search template: Even if the exact RNA sequence or structure is not yet known, analyzing essential and non-essential features of an RNA structure, for instance by affinity selection schemes, mutagenesis or genetic complementation, can accumulate sufficient data to yield a specific template to start with and look for in databases. This approach can be used to reveal related RNA structures in other genes or organisms even though a consensus template was not previously known. SELEX experiments provide a very versatile approach to derive RNA templates in this way (Türk 1997; see Chaps. 3 and 6; Fig. 3.1).
The folding of long RNA structures can be examined in detail to yield and predict specific regulatory elements in correlation with experimental data, for instance the RNA instability of different constructs. Such an approach is completely template-free and provides genuine regulatory RNA elements contained in the long RNA examined (see Chap. 6). Thermodynamic folding energies can also reveal biologically significant RNA energies if significantly lower than energies found in random sequences of the same length and/or if particularly low for one segment of the sequence considered in comparison to other segments of this sequence, yielding for instance novel regulatory motifs in viral 5' UTR (Konings et al. 1992).

Further, computer simulations of the dynamic RNA folding pathways are possible using genetic algorithms (Gultyaev et al. 1995).

Language-like approaches which, for instance, look for splice sites, introns and other more complex structures, are additional and quite powerful approaches to look for regulatory RNA structures in the absence of a template, provided they identify correct rules and context dependence of such varying nucleotide recognition sites. Thus, reading frames in long sequences can be better predicted by Fourier analysis (Tiwari et al. 1997). However, simple concepts, if intelligently implemented, are also quite powerful. Thus, identification of the correct AUG start codon for a reading frame can be greatly enhanced by looking out for a signal peptide in the correct position relative to it (see Chap. 6 for further rules for tackling this problem).

A complete search project is never trivial, the motif description requires careful study of the RNAs, the translation into a program has many alternative possibilities and different data banks may be available for the search. Each of these factors greatly influences the results obtainable. Another factor is even more important: none of the search programs is perfect, independent examination of the results by comparison with available literature and biological knowledge as well as a close interplay between theory and experimental testing of the best candidates are highly recommended. The reader should be further reminded of potential problems in extending his results to the human genome. It has been unclear how many model organism genomes will really be needed to delineate most hidden regulatory regions. Further information concerning such problems is discussed by Wassermann et al. (2000).

However, bearing these words of caution in mind, the reader is otherwise encouraged to start his/her own motif searches to reveal the full treasure of different motifs in the RNA molecules in which he/she is interested.

5. Functional RNA Interactions

5.1 Introduction

How do different RNA motifs exert their function, and which are their typical interacting partners? After the general description of the different RNAs and processes (Chap. 2) the focus of our book, regulatory RNA motifs, is reviewed in this chapter in the context of function and functional interactions.

5.2 Cellular Information Transfer from mRNA

From the primary gene transcript up to the defined translation of the mRNA reading frame into a protein, control steps and protein or RNA interactions mediated by RNA motifs regulate the release of the encoded mRNA information into the cell (see Fig. 1.1). Any type of higher differentiation depends on an increasingly selected release of information. Starting from the eukaryotic cell, the nuclear envelope prevents RNA precursors from passing into the cytoplasm. Instead, the precursors are processed into introns, heteronuclear RNA and mature RNA. Only mature mRNA passes into the cytoplasm. Also the DNA in the cell of a higher-developed organism contains the information for all tasks in the body, but only a very limited amount is transcribed and even less released in the cytoplasm and finally translated into proteins. The controlled release and, in fact, modulation of information release in the cytoplasm by various RNA structures and interacting protein factors has often been underestimated.

Many regulatory functions are fulfilled in the cell by RNA motifs. They are sometimes extremely fine-tuned for the exact readout, for instance a single methyl group may be critical for correct charging of tRNA (Pütz et al. 1994). Novel features for providing specificity in tRNA selection, including an amino-terminal domain, containing a novel protein fold that makes minor groove contacts with the tRNA acceptor stem, have been described by Sankaranarayanan et al. (1999). The specific release of information requires further detailed RNA interactions to control availability of the information (where in the cell? where in the organism?), and the resulting effect (protein synthesis, differentiation etc.).

Examples of the fine-tuned interactions are found in development. In general, the information may be sequestered in time (e.g. only available during development, during infection, during high concentration of metabolite x) or in space (e.g. only at the anterior pole, in the cytoplasm, in the nucleus). For each of these tasks different RNA signals have developed. Examples are: localization signals, such as in developmental genes in *Drosophila*; genes controlling the germ line development in *C. elegans* (Wickens et al. 1997, 2000); regulatory signals, as in the 3' UTR of U1A mRNA (Boelens et al. 1993); more are shown in Table 2.2.

The specific RNA elements in developmental RNA genes can be broadly divided into two functional classes according to whether they effect spatial or temporal release of information (see above and Chap. 3 for a number of RNA-protein interactions). In identifying new RNA signals for these dynamic processes, it must be borne in mind that both types of signals may be used in the organism under investigation. It is easy to overlook one or the other, as the experimental strategies to reveal them will be quite different, for instance *in situ* staining strategies to reveal localization patterns and signals for mRNA and protein expression time courses for the study of temporal regulation. The life cycle of an organism may nevertheless give some indication. In *Drosophila* the list of localization signals in regulatory RNA sequences is long because the very short time scale on which the fly develops effects selection for patterning genes in which spatial localization allows saving of time yet separate differentiation of the different body tissues. In contrast, organisms which develop over a longer time for, such as mammals, can in principle exploit time-dependent signals during development to a far greater extent. Thus, we suggest that in mammalian development there may be many RNA signals operating by time-dependent regulation of RNA activity, e.g. in mRNA translation or mRNA stability. The complex regulation of *c-fos* is a well-known example. mRNA decay in *c-fos* is regulated by AU-rich elements, found also in the 3' UTR of many other highly labile mammalian mRNAs. There are two distinct decay mechanisms, either processive but asynchronous ribonucleolytic digestion of poly(A) tails, or synchronous poly(A) tail shortening. Turnover of mRNA in yeast is promoted by the MATalpha1 instability element. The 65 nt segment of the MATalpha1 mRNA promotes rapid degradation of the mRNA (Caponigro and Parker 1996). This strengthens the hypothesis that deadenylation-dependent decapping at the 5'end is a common pathway of mRNA decay in yeast, and indicates that an instability element within the coding region of an mRNA can affect nucleolytic events that occur at both the 5' and the 3' ends of mRNA. Synchronous poly(A) tail shortening is the case for *c-fos* and implies a distributive or non-processive decay pathway (Shyu et al. 1989, 1991; Chen et al. 1995). Furthermore, additional signals are apparent though not yet understood; thus interleukin-3 mRNA stability is mediated by a signal in the 3' UTR independent of AU-rich elements, as shown by deletion analysis (Hirsch et al. 1995). An example of our own research is the human oestrogen receptor with its 3' UTR mRNA carrying an mRNA instability signal and its complex splicing, processing and translational control in the 5 'UTR, including short-sequence leader peptides (Kenealey et al. 1996).

However, in *Drosophila* many different regulatory pathways are explored for regulatory RNA motifs and interactions. Thus, *oskar* mRNA appears to be involved in two different circuits. Firstly, a gradient of *oskar* protein as a transcription factor is established *via* the localization of *oskar* mRNA to the posterior pole of the oocyte through an interaction which requires tropomyosin (Erdélyi et al. 1995). Secondly, besides concentration gradients, spatial patterns can also be established by different types of regulatory cascades . There is some indication that *oskar* mRNA is also involved in such a network (Breitwieser et al. 1996, see Chap. 6).

In *C. elegans tra-1*, a member of the GLI familiy is required for normal sexual development. It contains five zinc-finger domains and binds the identical DNA sequence (Jan et al. 1997); but *tra-1* can also act transcriptionally to govern gene activity. Binding of *tra-1* to the 3' UTR of *tra-2* mRNA regulates the export of *tra-2* mRNA from the nucleolus. The fact that *tra-1* is part of a conserved family of proteins raises the possibility that GLI members are both transcriptional and post-transcriptional regulators of gene expression (Graves et al. 1999). A method for controlling gene expression in a living cell has been described by Werstuck and Green 1998. Short RNA aptamers bind specifically to a wide variety of ligands. Insertion of such a small molecule aptamer into the 5' UTR of mRNA allows its translation to be repressible by ligand addition in vitro as well as in mammalian cells.

The use of mRNA stability signals for multiple regulatory effects is not limited to development. Dynamic aspects include decay regulation by hairpin stem loop-like structures such as iron-responsive elements (IREs) in the 3' UTR of TfR mRNA (Chap. 2 and 4) or processing regulation, which uses elements downstream of the polyadenylation site (Table 2.2). For instance, the translation of heat shock-specific mRNAs presents a very dynamic example (Chap. 3) where the different mRNA signals involved have not yet been sufficiently characterized. Another example illustrating the many instances where time-dependent regulation in the cell is mediated by regulatory RNA signals is the regulation of antifreeze proteins in winter flounder. The level of the type I antifreeze protein (AFP) protecting this fish against cold is regulated by its mRNA. Its level varies seasonally as much as 1000-fold. Duncker et al. (1995) have investigated the role of temperature on AFP accumulation using transgenic *D. melanogaster* by expressing multiple AFP genes under the control of the heat-inducible heat shock promoter. AFP and AFP mRNA persisted far longer in flies reared at 10°C compared to 22°C. This difference appears to be mediated by cold-specific mRNA stability since no such temperature effect was observed with either an endogenous heat-inducible mRNA or a constitutive expressed mRNA. It has been proposed that regulation of *D. melanogaster* neurogenesis might be regulated by RNA:RNA duplexes (Lai and Posakony 1998). The mRNA 3' UTR of Brd, hairy and other genes contain a novel class of sequence motif, GY box (GYB, GUCUUCC) and the 3' UTR of three proneural genes include a second type of sequence element, the proneural box (PB, AAUGGAAGACAAU). In the GYB and PB, the central 7 nts are exactly complementary, and are often located within extensive regions of RNA:RNA duplexes predicted to form between PB and GYB-containing 3' UTRs. It is suggested that in *D. melanogaster,* PB- and GYB-bearing transcripts may likewise participate in a regulatory mechanism mediated by RNA:RNA duplexes, but within the feature that both partners are mRNA that also directs the synthesis of functionally interacting proteins.

In any of the other steps of the pathway of RNA (Fig. 1.1), similar time-dependent functional interactions are observed. Thus, iron-dependent regulation in the 5' UTR by IRE*s* is effected in a time scale of hours in mammalian cells. In contrast, a very rapid (milliseconds to seconds) and only partly elucidated dynamic effect of RNA signals is the effect of RNA secondary structure on folding of the nascent protein chain during translation (Brunak and Engelbrecht, 1996; Thanaraj and Argos 1996).

5.3   Catalytic RNA

Catalytic RNA also takes part in the general task of controlled release of information. This includes direct information processing in the case of the dynamic events during splicing (Chap. 2, further details in Chap. 6). Catalysis is in most cases targeted to RNA, with the result that its information content is changed, for instance tRNA processing by RNAse P, or removal of intronic information from the mRNA precursor.

Pre-mRNA splicing generally occurs only in eukaryotes. Eubacteria very likely lost their introns, if they had any, as a selection for shorter duplication time of their genomic information; in Archaea some introns are left. In splicing, information is efficiently compartmentalized: usually introns stay in the nucleus and the exons are joined to give the mature mRNA. This compartmentalization depends further on the nuclear membrane and an elaborate transport process and should be noted as a further interactive process between different RNA and protein partners. Several regulatory RNA signals involved are discussed in Chapter 6.

The removal of introns by splicing of the precursor mRNA transcript to generate the mature mRNA occurs in the spliceosome in the nucleus. It has a parallel in rRNA processing, though the dynamic steps in the folding of the different rRNA intermediates are no doubt different, quite complex and only partly understood (see Chap. 6). It is interesting to speculate that the complex and evolutionarily ancient process of rRNA maturation may have provided useful processing activities, which have subsequently been mutated and are adapted for the mRNA splicing machinery. The main interaction partners with rRNA during its processing steps are the small nucleolar RNAs (snoRNAs) (together with attached proteins forming snoRNPs and nucleases; Chap. 6 gives details on snoRNA function and motifs). Interestingly, snoRNAs are found both in eukaryotes and in archaebacteria.

How efficient is catalysis by regulatory RNA and RNA elements? Comparing the efficiency of selected artificial ribozymes, for instance in *trans*-esterification reactions, it appears that RNA on its own can be engineered to be very versatile. However, its performance remains 2 to 4 orders of magnitude below the performance of protein-based enzymatic activities for analogous chemical reactions (Lohse and Szostak 1996). On the other hand, catalytic RNA reactions where RNA is acting in concert with attached proteins such as ribonucleoproteins in splicing and rRNA processing are powerful and fast. Thus, the whole rRNA is transcribed and processed within seconds, and tens of thousands of mRNA molecules are smoothly processed per cell every minute. Ribonucleoprotein particles (RNPs) are a powerful means of control of cell metabolism.

The figures of the splicing pathway, the spliceosome (Fig. 2.10) and different snRNA interactions (Figs. 2.11, 2.12) show several important regulatory RNA motifs in action: the native splicing machinery, termed the spliceosome, consists of the mRNA precursor, the small nuclear ribonucleoprotein particles (snRNPs) and auxiliary proteins. The principal component of a snRNP is a small RNA (snRNA). The key snRNAs for splicing, U1, U2, U4, U5, U6, form different snRNPs. In the case of U4 and U6, this is a common snRNP U4/U6 where U4 and U6 interact by a Y-shaped complementarity (Fig. 2.11) and are dynamically rearranged during splicing with U4 being removed and no longer inactivating U6 (Li and Brow 1996). Though the whole snRNP particle, for instance in U1 snRNA, has several proteins attached, the RNA motifs are involved in the central interactions of the whole splicing process. The snRNAs are the key partners of the splicing reaction itself and involved in the catalytic process (Madhani and Guthrie 1994; Newman 1994). Several RNA motifs indicate their intimate interaction (Figs. 2.11, 2.12). The snRNAs also provide an important scaffold for the auxiliary protein factors attached to them (Fig. 2.2).

Similarly, there is a specific pattern for snoRNAs if they are involved as methylation guide RNAs for ribosomal processing where, during the methylation process itself, there is a transient direct base-pairing interaction with the rRNA (Kiss-László et al. 1996; Fig. 2.16). Further RNA features again indicate interacting proteins such as those attached to the box C and D in the second class of snoRNAs (Fig. 2.15).

Regulatory signals in mRNAs are often involved in the readout of the message and influence its stability. Time scales vary greatly. Thus, a typical time course for IRE-dependent shut down of mRNA read out spans several hours, though in this more complicated process membrane permeability and cellular iron transport compartments are also involved in the regulatory loop. In contrast, the feedback loop for rRNA processing operates within seconds. *Trans*-esterification reactions occur very fast (10^{-9} s), however, there is no further regulatory control loop involved. In addition, one can investigate how different regulatory motifs, for instance in GCN4, regulate processes according to the requirements of metabolism (Hinnebusch 1997). A first example showed in the IRE that the response is directly dependent on cellular requirements: a positive 5' UTR regulatory loop utilizing a IRE positioned upstream of the open reading frame regulates synthesis of iron storage proteins such as ferritin (Hentze et al. 1987). Ferritin translation is blocked if the regulatory protein IRP binds and is released again by sufficiently high intracellular levels of iron. Another example (Dandekar et al. 1991) controls the first step of porphyrine synthesis, the mRNA for the delta aminolevulinic acid synthetase (eALAS), by an IRE element (Fig. 2.5). The synthesis pathway from aminolevulinic acid and glycine to the porphyrine ring is of little use in the cell if there is not sufficient iron present to synthesize functional haem groups with an iron ion in the centre. Furthermore, there is a negative regulatory loop operating at the transferrin receptor mRNA 3' UTR sequence using five IREs (Fig. 2.7), which leads to rapid decay of the message if no IRP protects the 3' UTR when the intracellular iron level is high.

The role of an upstream RNA structure in facilitating the catalytic fold of the genomic hepatitis delta virus (HDV) ribozyme was detected by Chadalavada et al. (2000). HDV has a circular genome that replicates by a double rolling-circle mechanism. The genomic and antigenomic versions of HDV contain a ribozyme that undergoes *cis*-cleavage. Pairing occurs at three sites. One site, termed Alt1, consists of an inhibitory stretch which forms a long-range pairing with the 3' strand of P2 located at the very 3' end of the ribozyme. Alt2 involves upstream nucleotide-ribozyme interactions and Alt3 involves ribozyme-ribozyme interactions. In other cases, the direct effect on metabolism is more complex. For example, we could show that the synthesis of two enzymes involved in the citrate cycle is controlled by IREs. By their position in the 5' UTR these elements suppress the translation of the mRNA for each of these enzymes under low levels of intracellular iron. A requirement for iron in these enzymes can be understood by the fact that these enzymes depend on iron-sulfur clusters to be catalytic. Another evolutionary advantage for these regulatory elements may be protection of the cell against oxidative side products of citrate cycle follow reactions (Gray et al. 1996). These are patients suffering from hereditary hyperferritinaemia and cataract. This is an autosomal dominant disorder characterized by a constitutive increased synthesis of L-ferritin in the absence of iron overload and associated with a point mutation in the IRE of L-ferritin mRNA. In the G51C mutant the binding of the mutant IRE to the IRP was reduced (Camaschella et al. 2000).

RNA is also specifically exported and imported to regulate information release and RNA metabolism. Different transport routes have been identified, and a picture is emerging that different transport cargoes use different transport vehicles (Görlich and Mattaj 1996). Nuclear protein import occurs using specific protein sequences (nuclear localization signals, NLSs). For RNA export different RNA signals are recognized: a cap-binding complex recognizes the monomethylated 5' end of snRNAs and mRNAs. An increasing multitude of protein components involved in RNA transport is being revealed (Doye and Hurt 1995; Görlich and Mattaj 1996) and is involved, particularly in RNA export from the nucleus.

The nucleoporins include several proteins, which recognize RNA; however, other RNA-binding proteins do not always localize to the nuclear pore complex. Furthermore, their exact signals, as well as the relevant RNA signatures, have to be identified, for instance the nucleoporin Mex67p binds to poly(A) -stretches (Ségref et al. 1997).

5.4   RNA Motif Evolution

As with every other living entity, evolution operates on RNA sequences and RNA structure. Comparison of different RNA species in evolutionary trees has a long tradition (reviewed in Gesteland et al. 1999). Furthermore, at least for RNA secondary structures, more comprehensive theoretical and evolutionary treatments exist. Thus, neutral paths have been described to change evolutionarily from one RNA structure to another by successive mutations enabling circumvention of high-energy barriers (Schuster et al. 1997). Though these are simplified treatments ignoring the additional complications of tertiary interactions, RNA secondary structure is inherently more locally than protein structure and thus such simplified evolutionary path predictions are more feasible for RNA than for proteins. The significance of calculating and mapping neutral paths, although at present only applicable to simple RNA structures, is firstly that such paths allow in principle a comprehensive, complete treatment of all evolutionary changes and sketch the way from one structure (or one function) to another. The same goal for protein structures is at present much less feasible. Apart from this powerful theoretical justification, which unfortunately has not yet received the attention it deserves, neutral paths and the map of surrounding mutations are also important for delineating the stability of different RNA structures and motifs, to understand their functional flexibility and essential and non-essential paths. A third implication, which may soon be better appreciated, in particular by experimentalists, is that the study of neutral paths and structure flipping via such paths is a powerful tool to sketch or estimate the success of SELEX experiments (Burke and Gold 1997). SELEX experiments can be tuned to yield RNA molecules with multiple adaptations or aimed at changing an RNA structure to adopt additional desired features.

At least for smaller three-dimensional RNA structure and regulatory motifs, surprisingly exhaustive studies regarding mutation and flexibility are possible.

To encourage the reader to investigate his or her own RNA regulatory element by similar approaches, we mention the results by Olsthoorn et al. (1994, 1996). This group investigated how RNA helices are selected and survive during evolution. One example, the mRNA for RNA replicase in different phages examined (Fig. 2.17) shows interestingly that both the RNA structure (a stem loop structure at the 5' end of the open reading frame) and the amino acid codons (Tyr-Gly-Asp-Asp) are conserved, though the register has not been exactly retained. To retain at the same time several conserved amino acid positions as well as the RNA stem loop requires among other things a viable evolutionary path with good fitness and survival of each mutated variant of the phage. In the fitness landscape of different RNA mutations the RNA motif (the stem loop) could be kept though its position within the mRNA had changed. To shift the RNA motif, this evolutionary path probably had to be achieved via mutations, which were neutral for the encoded protein sequence. Moreover, this reveals also more details about the regulatory motif: this is another example that at least such evolutionarily maintained RNA structures may be more involved in translation than previously thought.

5.5   RNA-Protein Interactions

Many RNA-RNA and RNA-protein interactions co-evolve, as is shown by compensatory base pair changes, for instance in the U4/U6 interaction. However, a wide range of interactions can easily evolve by duplication of RNA elements and transfer of the RNA position, possibly together with acquired mutations. The different regulation for the IRE motif depending on its position in the 5' or 3' UTR of mRNA is clear evolutionary evidence for this. Another example is the shared snRNP proteins where some evolved to snRNA specific variants. This book has already covered many examples of RNA-protein interactions. For example, RNA sequence-specific regulatory proteins typified by the rev-protein from HIV prevent splicing of the message by fast transport of the RNA out of the nucleus. Table 5.1 summarizes several important examples, which together with those given in Table 2.4 briefly summarize protein interactions, which can be guided by RNA structure. RNA elements important for protein interactions are provided in Table 2.2. There are more known instances for each RNA-protein interaction mentioned; thus, in several of the developmental examples cited protein partners are involved (e.g. the 3' UTR of *staufen* or *bicoid*). Similarly, many of the examples of viral regulation or 5' UTR regulation in bacteria and eukaryotes given in Table 2.2 depend on interacting proteins.

Table 5.1. Examples for RNA structure specific protein interactions

Example	RNA structure	Specific protein(s), interaction
Viral life cycle		
HIV	RRE	rev protein (Cullen 1994)
	TAR	Tat protein (Cullen 1994)
Metabolic control		
GCN4	5' UTR	GCN proteins, regulatory motifs: short upstream ORFs (Hinnebusch 1997)
Transferrin receptor	3' UTR IRE	IRP (Rouault et al. 1989)
Ferritin	5' UTR IRE	IRP (Hentze et al. 1987, new examples: Chap. 6; IRP1, IRP 2 (Weiss et al. 1997)

Autoregulatory interactions

Ribosomal protein	mRNA	Encoded ribosomal protein (see Chap. 2)
Histone protein	mRNA	Encoded protein(Peltz and Ross 1987)
Per Gene	mRNA	Encoded protein (Hardin et al. 1990)
Beta-tubulin	mRNA	Encoded protein (Gay et al. 1989)
Human pre-pro-insulin gene	mRNA	Encoded protein (Knight and Dochetry, 1991)
U1A	mRNA	Encoded protein (Boelens *et al.*, 1993)
Ribonucleoprotein complexes (RNPs)		
ribosome	rRNA	Ribosomal proteins, rRNA as scaffold
RNAse	RNAse P	RNA as protein scaffold
Translocation	7SL RNA	SRP proteins
Splicing	snRNAs	snRNP proteins

RNA-protein interactions between RNA motif and binding protein also trigger complex differentiation processes such as sex determination. For example, pre-mRNA splicing of the *transformer* pre-mRNA in *Drosophila* depends on binding of the sex-lethal protein (sxl), present only in female flies. sxl binds to a non-sex-specific splice site and diverts the splicing factor U2AF from it, so that a more distal 3' splice site is used for splicing in females. In the resulting mRNA a stop codon is skipped and functional transformer protein synthesized in females. In male flies the normal splice site is used and the stop codon prevents synthesis of a functional transformer protein (Sanchez *et al.* 1994; Kelley and Kuroda 1995). Splice site selection is often determined by competition between alternative binding sites for the different splicing factors (Chaps. 2 and 6), and further tissue or differentiation specific protein factors. Several alternative splicing examples are listed in Fig. 2.13; *Drosophila double-sex* (Steinmann-Zwicky 1994) is a further sex-specific example. Another example is represented by bacteriophage T7 lysozyme, which binds to T7 RNA polymerase and regulates its transcription by differentially repressing initiation from different T7 promoters, destabilizes initial transcription complexes and increases the rate of release from theses complexes. The resistance of the elongation complex to T7 lysozyme seems to be due to the consecutive establishment of two sets of RNA:RNP interactions Huang et al. (1999). In P12 cells, the RNA-binding protein HuD plays a critical role in PKC-mediated neurite outgrowth by primarily promoting the stabilization of the GAP-43 mRNA (Mobarak et al, 2000).

Modified Nucleotides in RNA

During widely accepted but still hypothetical RNA world stage of life (Gesteland et al. 1999), RNA had to fulfill even more functions than now. Thus, there was a high evolutionary pressure to develop suitable chemical modifications of its structure for broadened functions and interactions. It is thought that the nucleotides constituting RNA were scarce on earth. RNA-based life must therefore have acquired the ability to synthesize RNA nucleotides from simpler and more readily available precursors, such as sugars and bases. The main problem was to find a way of coupling sugars, sugar phosphates and the RNA bases together forming pyrimidine nucleotides. Unrau and Bartel (1998) further supported the RNA world hypothesis by the discovery that RNA can even catalyze the synthesis of pyrimidine nucleotide at their 3' terminus. For further analysis of the result see also Koch and Dandekar (1999). An alternative view is that in present-day competition with proteins requires modification of RNAs to allow RNA to retain functions where it would otherwise be outperformed by proteins. Both views seem to converge in the sense that in those reactions where RNA is the leading molecule (for instance ribosomal RNA processing and the different tRNAs) most modified nucleotide adaptations can be found. In addition, modifications such as the 2,2,7-tri-methyl-guanosine cap in small nuclear RNAs or the mono-methyl-guanosine cap of mature mRNA (m^7Gppp-cap) show that also here modified nucleotides provide a first signal for proper recognition, transport or processing of the RNA in question. The m^7Gppp-cap as a 5' modification is acquired by all RNA polymerase II transcripts and is thus present in both mRNA and freshly synthesized new snRNAs (apart from snRNA U6, a RNA polymerase III transcript). This regulatory signal leads also to the export of newly synthesized snRNAs into the cytopasm. There, they are assembled with proteins to form the complete small nuclear ribonucleoprotein particles snRNP U1, U2, U4, U5 and U6 and the cap is modified to carry the 2,2,7-tri-methyl-guanosine signal which is recognized in the transport of the assembled snRNP back into the nucleus (Görlich and Mattaj 1996).

Identification of Modified Nucleotides

Detection of modified RNAs is a challenge. In fact, the amount of RNA modifications is probably still underestimated, simply because so many RNA sequences are only deduced from the DNA sequence and even when more directly determined, there is often not a direct evaluation of modification. Most of the examples of modified nucleotides known to date were only found after looking in more detail at the function of specific RNA metabolism regulation. Furthermore it should be pointed out, that standard database searches (see Chapter 4) normally ignore modified nucleotides, and to further complicate matters, RNA sequences in which modified nucleotides are indicated by special characters may even bring a badly designed program to crash - or will be simply ignored. On the other hand, if the surrounding nucleotide features are known and properly described as a motif a specific search for such motifs is possible. The modifications for RNAs are astonishingly specific and, in this sense, constitute micromotifs of just one modified nucleotide, e.g. the snRNA cap. The detection is simplified, if one is aware of this possibility and also able to look for it. This can be done either by a suitable motif description of surrounding nucleotides involved in appropriately guiding the modification of the RNA (using the techniques outlined in Chap. 4), or by experiments targeted to identify modified nucleotides.

Modified Oligonucleotides as tools to Modulate RNA Metabolism

Chemical modifications also allow modified RNAs to be used specifically as tools, for instance in antisense technology or as more stable probes. They are very useful for identifying or verifying suspected regulatory motifs in RNA. RNA sites may be protected from enzymatic cleavage and confirmed in this way. Furthermore, RNA-RNA interactions are revealed, by making central interaction areas more resistant to cleavage. In addition, detection or affinity purification of certain RNA motifs is enabled, by attaching and using modified nucleotide groups. Table 5.2 gives a summary of several currently used chemical modifications, Johansson et al. (1993) and Pütz et al. (1994) describe more examples. Further practical implications are: Probes can be made more stable or can be specifically used to cut RNA when forming hybrids by endogenous or additional administered double strand specific RNAs and in this way be used to target and cut identified regulatory RNA motifs.

Table 5.2. Modified oligonucleotides as tools

Chemical modification	Property/application
2-O-Allyl	More stable
2-O-Methyl	More stable
Biotinilation	Affinity purification
Thiolate groups	Enzymatic blocking

Interaction Dynamics

A very general way to phrase the dynamics of regulatory RNA and RNA elements is to describe it in terms of patterns in space and time. The numerous developmental interactions where RNA is involved (Table 2.2 and Chap. 3) are evidence for this. Short-term and dynamic interactions involve catalytic and processing steps such as those, which occur during splicing and within the spliceosome (Madhani and Guthrie 1994). The pre-splicing complex forms within minutes *in vitro* and probably orders of magnitude faster in the cell. Similarly, ribosomal processing occurs rapidly within seconds in yeast cells (Venema and Tollervey 1995). We point out that these time scales apply to the whole processing. The single molecular interactions are even faster- and the only chance of identifying them is normally analyzing the RNA motifs involved for complementarity and potential interactions (see Figs. 2.11, 2.12 for a summary of different spliceosomal interactions involving U1, U2, U4 and U6). After being identified by virtue of the different RNA motifs contained, firm proof can only come from experiments involving mutagenesis and compensatory base-pair interactions, e.g. regarding the U4/U6 interaction (Li and Brow 1996).

Another aspect of the dynamics of RNA-protein interactions is how RNA structure influences protein folding. Recent examination of pre-mRNA processing by proteins containing the 70-90 amino acid alpha-beta module known as the ribonucleoprotein domain showed that the specificity of ribonucleoprotein interaction is determined by the folding of the RNA during the complex formation (Allain et al. 1996). The human U1A protein binds the RNA hairpin of snRNA U1 during splicing, but also regulates its own expression by binding an internal loop in the 3' UTR of its pre-mRNA, preventing polyadenylation (Boelens et al. 1993). The NMR structure of the complex between the regulatory element of the U1A 3' UTR and the U1A protein RNA-binding domain revealed that it requires the interaction of the variable loops of the ribonucleoportein domain with the well-structured helical regions of the RNA. The forming complex next orders the flexible RNA single-stranded loop against the protein beta-sheet surface and finally reorganizes the carboxy-terminal region of the protein to maximize surface complementarity and functional group recognition (Allain et al. 1996).

Co-translational mRNA structure may also be important, both for the dynamics of the mRNA translation by the ribosome as well as protein folding of the nascent amino acid chain, Brunak and Engelbrecht (1996) stress the first point. The authors use neural networks based on a complete set of 719 protein chains (with high and low expression) and known three-dimensional structure. The corresponding mRNA was used to identify codons for very abundant amino acid residues. In the encoded protein with known structure these were the amino acid positioned in N- and C-termini of helices and sheets. No signal could be found for rare codons. However, the mRNA signals could be compared with conserved nucleotide features in the 16S rRNA sequences and related to mechanisms for the maintenance of the correct reading frame by the ribosome. Thanaraj and Argos (1996) focus on highly expressed proteins from *E.coli* and take codon frequency (frequent codons are also translated by more abundant tRNA species) as an indication for translational speed. They find that translationally fast mRNA regions tend to encode helices while slow regions prefer sheet and coil. They further observe absence of strong positive selection for codons in non-highly expressed genes, compatible with existing theories that mutation pressure may dominate codon selection in the non-highly expressed genes.

Ribosomal RNA biogenesis is a procedure in which nucleolin, a 70 kDa nucleolar protein is involved in many ways. The two most N-terminal domains RBD1 and RBD2 of nucleolin bind to an RNA stem loop containing the sequence UCCCGA in the loop. Both RDBs interact with the RNA loop via their beta sheet. Each domain binds residues on one side of the loop, RBD2 contacts the 5' site and RDB1 contacts the 3' site (Allain et al. 2000).

Ribozymes are simpler to analyze. Dynamic interactions involve complementarity within the same molecule itself, e.g. in the hammerheads and hairpins shown in

Figs. 2.21 and 2.22. Again, the description of the motifs involved is critical in delineating the reaction scheme, which afterwards is tested by deletion and mutagenesis experiments. The time scale is even faster (as no additional interaction partner is required) and has almost the speed of proteinogenic enzymatic reactions (microsecond time scale).

RNA structures have many possibilities to derive catalytic potential, the RNAse P cage (Fig. 2.23) and the structures of different self-splicing introns (Figs. 2.25, 2.26) show some further possibilities. According to Schuster et al. (1997) for most catalytic structures there are enough possibilities for neutral mutations, which do not interfere with their catalytic activity that the structure may ultimately flip with a final, non-conservative mutation into a new catalytic activity. In fact, re-engineering of critical RNA motifs in catalytic RNA structures to achieve novel properties such as those, shown in Fig. 5.1 (from Sullenger and Cech 1994) is a worthwhile exercise (Jones et al. 1996). The primase of *E. coli* is required for the production of primers to initiate DNA replication. Its crystal structure has been determined comprising the *E. coli* DnaG protein showing that DnaG binds nucleic acid in a groove clustered with invariant residues and that DnaG is positioned within the replisome to accept single-stranded DNA directly from the replicative helicase (Keck et al. 2000).

Basic principles are conserved in many dynamic interactions. A well-known example described in Fig. 2.10 compares similar features in the splicing of group I introns and in the far more complex spliceosome with its many RNA interaction steps (Cech 1993). In both cases the reaction proceeds via an intermediate to the splice site exchange and subsequently yields the spliced RNA. Similar reactions are apparent in the other types of splicing such as group II self-splicing introns (Fig. 2.26), *trans*-splicing (Fig. 2.14) or ATAC-intron splicing (Fig. 6.1). Group II self-splicing introns (Fig. 2.26) are compared to group I self-splicing in Fig. 2.25 and the overall similarity in the pathway is evident. However, the group II intron as a consequence of its different RNA structure (Fig. 2.26) shows a branchpoint and a branched intermediate as an important structural difference during the dynamics of the splicing reaction. Spliceosomal complexity (reviewed in Lamond 1995) allows fine-tuning of pre-mRNA processing and pre-mRNA activation for processing, e.g. in different tissues or developmental stages. Furthermore, the many factors involved in mRNA splicing allow high efficacy, accuracy and speed of this dynamic mechanism for fast mRNA processing.

5.6   Designing Interactions

The range of reactions for which ribozymes have catalytic capabilities is quite large. Template-free synthesis of longer RNA molecules has still not been achieved (first steps in this direction are reviewed in Koch and Dandekar 1999). This would be the most compelling proof for the ability of RNA to exist and multiply on its own. However, relying on unmodified, pure RNA may be demanding more from the system than was originally required in nature. Chemical modifications, naturally occurring modified nucleotides (see Table 2.5) and additional prosthetic groups are now omnipresent as cofactors in many enzymes and may have helped in the action of the premordial RNA polymerase.

The power of RNA to recognize virtually any substrate is underlined by SELEX approaches allowing isolation of RNA species with binding affinities and K_m between 10^{-9} to 10^{-4} M for substrates as diverse as chromophores, proteins and metal ion complexes (Lohse and Szostak 1996). The frontiers of RNA evolution may perhaps be more finely delineated in the future by direct comparison to SELEX based peptide evolution (see Chap. 3).

Many of the rapidly growing number of genome sequences analyzed today encode for active RNAs which still have to be discovered, one of the main incentives for writing this book. Obvious examples for new molecules to be found include *trans-splicing* RNAs, snoRNAs and IREs. We have witnessed a constantly growing list of new candidates, from which several could ultimately be directly experimentally verified as carrying the function and structure assigned by identifying the RNA motifs. This is probably only a tiny list of further possible regulatory RNA types. There is a growing list of catalytic motifs (Table 2.2) and similarly more and more modulating motifs in the 5' and 3' UTR of mRNA are being identified and are currently being analyzed in different laboratories. In particular the mRNA examples show that regulation can be quite complex involving different spliced mRNAs, regulation by very short upstream ORFs in the 5' UTR, RNA editing and different stability signals in the 3' UTR (Jackson and Wickens 1997; Wickens et al. 1997; Wickens et al., 2000). Evidence is accumulating that for many of the more interesting regulatory mRNAs such a complex scenario of regulation is rather the rule than the exception. This is a region of research which has previously been rather neglected due to a tendency to concentrate on the ORF and the encoded protein, but which is now rapidly catching up.

Another trend is that RNA opens up ever more different levels of regulation in the cell. In the nucleus this is not only limited to the different types of splicing, intron processing and ribosomal processing in the nucleolus. Additional signals include various transport signals such as the RRE and differential splicing signals, for instance in sex-specific splicing. The metabolism of the heterogeneous nuclear RNA also gives new surprises, including more new functions for the spliced introns.

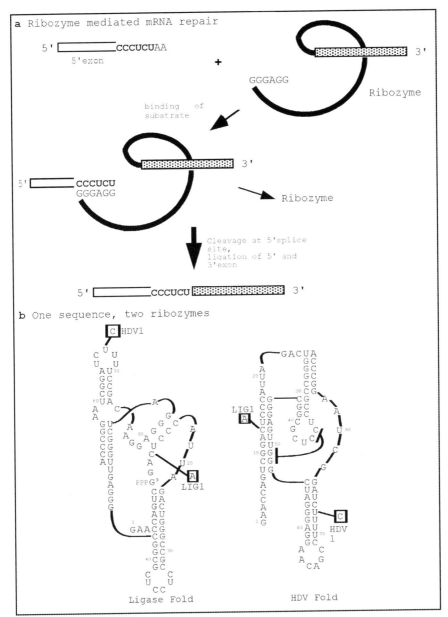

Fig. 5.1ab. Ribozyme-mediated repair of mRNA. Depicted is the principle of mRNA repair by ribozymes according to Sullenger and Cech (1994) and Jones et al. (1996). **a** A missing 3' exon or a "repaired" exon can be provided to the 5' exon via *trans*-splicing from the ribozyme (5' splice site sequence is indicated in *bold*) **b** An intersection sequence is shown threaded through both the class III ligase and HDV ribozyme secondary structures. The single-nucleotide substitutions that generate constructs LIG1 and HDV1 are indicated. (See also Schultes and Bartel 2000)

The recent discovery of many small nucleolar RNAs encoded in different introns (Seraphin 1993; Kiss-László et al. 1996; Ni et al. 1997) e.g. the two small nucleolar RNAs U14 encoded in *Xenopus* introns, or the intron encoded U21 in vertebrates and *Drosophila* is a step in this direction. However, phenomena such as tissue specificity mediated by introns, e.g. in rat fatty acid synthase (Oskonian et al. 1997) await further investigations and explanations.

Cytoplasmic regulation steps involve post-transcriptional regulation such as mRNA stability elements and different regulatory elements in the 5' UTR; apart from metabolically controlled elements this can be different short ORFs, stem-loop structures and complementarity regions to antisense RNAs. However, events such as sorting to the endoplasmatic reticulum also involve different types of regulatory RNA, such as the 7S RNA and the signal recognition particle. Again we point out that several more cytoplasmic RNAs exist as abundant species but their function is not always known. Thus, the function of 10Sa RNA in error correction came as a surprise, but the RNA band had already been apparent for some time in cytoplasmic Northern analysis. Further possible regulatory elements are different localization signals in developmental RNAs and at protein interaction sites. It is obvious that there will be many further instances in this field where the function of new regulatory elements will be described.

In biotechnology and *in vitro* applications, the use of RNA as a tool for the design of RNA technology has many surprises and new applications. A case in point is the design of nanopatterns: particles can be coated by different oligonucleotides. Such particles assemble specifically according to the oligonucleotide pairing capabilities. One method (Mirkin et al. 1996) involves assembly of colloidal gold nanoparticles by attaching to the surface two batches of 13 nm gold particles non-complementary DNA oligonucleotides capped with thiol groups, which bind to gold. Upon adding to the solution an oligonucleotide duplex with sticky ends that are complementary to the two surface-attached sequences, the nano-particles self-assemble into aggregates. The process is reversible by thermal denaturation. The strategy could, of course, be applied to similar tasks involving rearrangement on a nano-size scale of components using attached different DNA, RNA and chemically modified oligonucleotides.

5.7   Medical Implications

Curing RNA by *trans*-Splicing

Great interest has been aroused by ribozyme-mediated *trans*-splicing reactions as they have a great potential for use as therapeutic agents (Cech 1988; Castanotto et al. 1994). Ribozymes can be used to replace defective portions of a target RNA by functional sequences, as shown by Sullenger and Cech (1994), in which a group-I intron was designed to replace truncated lacZ transcripts by proper coding regions (Fig. 5.1). *Trans*-splicing ribozymes may also find application as antiviral agents, as one can change the sequence of the target viral RNA so that it encodes a dominant negative version of the protein (Malim et al. 1989; Trono et al. 1989). The implication of this finding is that any variety of mutant RNA associated with a variety of genetic diseases (e.g. cystic fibrosis and sickle cell anemia) could be repaired by employing such a strategy. The advantage would be that the gene could be retained in its natural environment to allow its expression to occur unaltered (Sarver and Cairns 1996). Though a *trans*-plicing cure of mRNA seems to be academic, *in vivo* repair of modified or pathologically mutated RNAs is quite feasible if efficient adenovrial delivery systems containing the curing RNA with its *trans*-spliced leader are developed.

Jones et al. (1996) developed a technique based on a tagged ribozyme to mediate *trans*-splicing in mammalian cells. However, the specificity of the reaction was low as the tagged ribozyme was shown to react with additional cellular transcripts other than the specifically targeted. This points to a second obstacle: besides delivery of the *trans*-splicing RNA, there may be adverse effects by inaccuracies of the reaction, which may modify mRNAs in dangerous, e.g. oncogenic ways.

Antisense Approaches

This is one of the areas where RNA is superior in many instances in the cell. Therapeutic measures might aim to control misregulation on the RNA level. This would involve, for instance, modulation of IREs to treat disregulation of iron metabolism. Though in many cases with clinical relevance direct substitution of iron is sufficient as in many types of anemia, more complex regulatory deficiencies would profit greatly from therapeutic efforts aiming at the RNA regulation via the IREs. These include hemochromatosis, where a pathologic overincorporation of iron might be countered by inhibition of transferrin- or transferrin receptor synthesis achieved by modulation of IREs involved. A well-established pharmacological modulator of red cell production, erythropoietin, has recently been shown to enhance the affinity of IRP-1 protein binding to the stem-loop structures of IREs and thus increase cellular iron uptake as more transferrin receptor is produced from its stabilized mRNA (Weiss et al. 1997). In porphyria, there is increased synthesis of intermediates in the delta-aminolevulinic acid pathway normally leading to hem groups, because some enzymes are not functioning properly. By influencing the first-step enzyme through the modulation of its synthesis by inhibition of its IRE, this illness could be alleviated

Further possibilities for the inhibition of unwanted mRNA include endogenous RNAse H-mediated cleavage of the RNA targeted by short, complementary oligonucleotides due to the formation of a double-stranded RNA hybrid.

As described for IREs, any of the regulatory RNA elements identified so far are amenable to similar therapeutic interference. For a number of *in vitro* translation systems, the efficacy of antisense oligonucleotides in suppressing expression could be clearly demonstrated. The same is true for the transfection of antisense constructs. The current and long-standing challenge is appropriate delivery of the antisense construct to a sufficient number of cells *in vivo*. However, also in this regard there has been constant progress. A first successful example is the inhibition of intima growth after arterial lesion by blocking growth using anti-myb antisense nucleotides (Simons et al. 1992). Therapeutic antisense nucleotides can, however, be employed not only as oligonucleotides but also as expressed nucleotides (Askari and McDonnel 1996). Expressed nucleotides are transcribed from an expression vector used for gene therapy in the patient, such as an adenoviral, retroviral or plasmid vector. However, such an approach has to deal with issues of safety and efficacy.

Antisense therapy against HIV is currently being attempted by targeting antisense nucleotides against the proviral long terminal repeats or regulatory elements of HIV expression such as transactivation response element (TAR) and rev-response element (RRE). Further areas of antisense therapeutic efforts are hypertension, where antisense oligonucleotides against the mRNA for angiotensinogen are infused into the liver or brain of hypertensive rats, resulting in a decrease in angiotensinogen levels (Tomita et al. 1995). Based on similar principles, anticancer treatment by blocking oncogenic messengers is being investigated (Ratajczak et al. 1992). A recent example of successful oligonucleotide antisense therapy is the work of Colomer et al. (1994). ErbB-2 is an oncogene, which is involved in the cascade leading to the proliferation of cancer cells in breast carcinomas. The mRNA for erbB-2 was inhibited successfully by applying antisense oligonucleotides. The direct inhibition of c-erbB-2 opens up new alleys for cancer therapy, stopping its pathological overexpression in epithelial ovarian cancer (Felip et al. 1995). A future aim of such approaches is to influence the complicated network of apoptosis and cell cycle regulation at the RNA level, for instance by inhibition of bcl-2 mRNA (Hockenbery 1995).

Problems with antisense therapy include the destruction of the oligonucleotides by DNAse or RNAse, which are ubiquitous in the body. Furthermore, these approaches are complicated by the large doses required for therapeutic effects or cell penetration (Clark et al. 1995) and difficulties in specifically directing an oligonucleotide to a certain type of cells (instead of inhibiting a central mRNA everywhere where the oligonucleotides penetrates a cell). In direct cell culture, repair of thalassemic human beta-globin mRNA by antisense nucleotides has been achieved (Sierakowska et al. 1996). However, antisense oligonucleotides have to be admistered by infusion (parenteral), and without chemical modifications the half-life of oligonucleotides in the plasma is short. Strategies to increase duration include replacement of the phosphodiester backbone by phosphorothionate, attaching oligonucleotides to DNA-protein complexes or to cationic liposomes. Directing the oligonucleotide can be attempted by use of cell-specific antibodies or receptor ligands. Thus, oligonucleotides complexed with asialo-oroso-mucoid can be targeted to the liver with improved selectivity (Lu et al. 1994). This example may be misleading, however, as the principal barrier for targeting drugs to places other than the liver are liver-specific cells, in particular the reticulo-endothelial cells, which try to eliminate all alien compounds and complexes including drug delivery complexes from the bloodstream. For instance, fusigenic

viral liposomes are a promising alternative which can be targeted to cardiovascular cells (Dzau et al. 1996).

New therapeutic routes are being examined, for instance intravitreal (into the eye-
ball) injections for cytomegalovirus retinis (a serious infection of the eye and the
retina) in HIV patients. Asthma in the mite-conditioned allergic rabbit model can
be treated by phosphorothioate antisense oligodeoxynucleotides targeting the
adenosine A1 receptor mRNA by an aerosole (Nyce and Metzger 1997)

Long-term effects of antisense therapy have still to be investigated, but
nevertheless direct inhibition of RNA is a safer and more precisely controlled
alternative than aiming for the same therapeutic effect on the DNA level by use of
vectors for gene modification. Another possibility to mention is that antisense
therapy with oligonucleotides can be allele- or subfamily-specific to fight disease-
specific alleles or even a whole subfamily of disease-specific genes.

Antisense approaches are also a powerful tool in clinical and basic research to
verify disease targets (Narayanan and Akhtar 1996). Thus inhibition of
tumorigenesis (inhibition of anchorage independent growth in adrenocortical
carcinoma cells) has been achieved using a cytosine-DNA methyltransferase
antisense oligodeoxynucleotide (Ramchandani et al. 1997).

RNA processing is also important for cellular immortalization, for instance
Malyankar et al. (1994) showed that the protein/proliferin transcript is degraded
by cancer cell-specific nuclear factors. A more direct medical interference with
such processes would be highly desirable. Moxham et al. (1993) showed that
induction of G-alpha-i2-specific antisense RNA *in vivo* inhibits neonatal growth in
transgenic mice and inhibits the action of insulin in cells (Moxham and Malbon
1996). Induction of antisense RNA of a fragment of the *S. aureus* alpha-toxin gene
(hla) led to a reduction in alpha-toxin expression in two different murine models
of *S. aureus* infection and reduced the lethality of the infection completely. This
shows that it is possible to eliminate alpha-toxin virulence in *S. aureus* infection
(Ji et al. 1999).

Modulating mRNA Stability

Tumor necrosis factor-alpha (TNF-alpha) is a major mediator of acute and chronic
inflammatory responses in many diseases. Tris-tetraprolin (TTP), a Cys-Cys-Cys-
His (CCCH) zinc-finger protein, destabilizes TNF-alpha mRNA. This is mediated
by direct binding of TTP to the AU-rich element. TTP is a cytosolic protein. Its
biosynthesis is induced by the same agents that stimulate TNF-alpha production; it
can be seen as a component of a negative feedback loop concerning TNF-alpha
production. This pathway might be a potential target for anti-TNF-alpha therapies
(Carballo et al. 1998)

HIV-1 transcription elongation is regulated through binding of the viral Tat
protein to the viral TAR RNA stem-loop structure. An 87 kDa cyclin C-related
protein (cyclin T) that interacts with the *trans*-activation domain of Tat is a partner
for CDK9, an RNAPII transcription elongation factor. The interaction of Tat with

cyclin T enhances the affinity and specificity of the Tat:TAR interaction. Tat might direct cyclin T-CDK9 to RNAPII through cooperative binding to TAR RNA (Wei et al. 1998). Tat induces an RNA-binding protein in central nervous system cells that associate with the viral *trans*-acting-response regulatory motif. It is thought that NF-kappa B p50 is such a protein and that Tat-mediated *trans*-activation of cellular as well as viral genes might contribute to the central nervous system damage associated with HIV-1 infection (Kundu et al. 1999). Insights into the mechanisms of viral genome recognition for extensive amino acid conservation of nucleocapsid protein (NC) of HIV-1 are provided (De Guzman et al. 1998). The three-dimensional structure of HIV-1 NC bound to the SL3 stem-loop recognition element of the genomic Psi RNA packaging signal has been revealed by applying magnetic resonance spectroscopy. A specific interaction between the amino- and carboxy terminal CCHC-type zinc knuckles of NC and the G6-G7-A8-A9 RNA tetraloop has been shown (De Guzman et al. 1998). The detailed structural analysis offers a basis for the development of inhibitors designed to interfere with genome encapsidation.

Group II Introns as an Engineering Tool

Another interesting point is the insertion of group II introns into therapeutically relevant DNA target sites in human cells. Mobile group II intron RNAs insert directly into DNA target sites and are then reverse-transcribed into genomic DNA by the associated intron-encoded protein. Guo et al. (2000) have developed a genetic assay in *E. coli* to determine detailed target site recognition rules for the *Lactococcus lactis* group II intron L1 and to select introns that insert into desired target sites. Using HIV-1 proviral DNA and the human CCR5 gene as examples, they showed that group II introns can be re-targeted to insert efficiently into virtually any target DNA and that the re-targeted introns retain activity in human cells. This might be the practical basis for applications in genetic engineering, functional genomics or gene therapy.

Ribozymes

Ribozymes are larger than oligonucleotides, and hence potentially more difficult to admister. However, a ribozyme molecule is catalytic and can modify many RNA molecules. It needs no host machinery to cut (though hybrid formation by the antisense oligonucleotide is often sufficient for inhibition and a cut by double stranded RNA cleavage enzymes provided by the host is not necessary) and can be very specific for one mRNA, furthermore they can block dominant mRNAs in genetic diseases. Besides viral diseases and cancer treatment, dominant genetic disorders are a principal area where ribozymes may be helpful, for instance in connective tissue disorders such as osteogenesis imperfecta, Marfan syndrome and the craniosynostotic syndromes (Grassi and Marini 1996. Lewin and Hauswirth 2001).

A single RNA sequence that can assume either of two ribozyme folds and catalyze the two respective reactions has been discovered (Schultes and Bartel 2000). The two folds share no evolutionary history and are completely different, with no base pairs, and probably no hydrogen bonds in common. Minor variants of this sequence are highly active for one or the other reaction. Thus, in the course of evolution, new RNA folds could arise from the preexisting fold without the need to carry inactive intermediate sequences. The authors draw the conclusion that RNAs having no structural or functional similarity could share a common ancestry and even functional and structural divergence might even precede rather than follow gene duplication.

The self-cleaving ribozyme of the hepatitis delta virus (HDV) is the only catalytic RNA known to be required for the viability of a human pathogen. Ferre-D'Amare et al. (1998b) revealed the 2.3 Å resolution structure of a 72-nucleotide, self-cleaved form of HDV ribozyme binding a small, basic protein without affecting ribozyme activity by X-ray diffraction showing a compact core that comprises five helical segments connected as an intricate nested double pseudoknot. (The 5'-hydroxyl leaving group resulting from self-scission reaction is buried deep within an active-side cleft). The structure information might help to open up therapeutic strategies against HDV.

Another ribozyme with potential therapeutic relevance is the hairpin ribozyme, which belongs to a family of small catalytic RNAs that cleave RNA substrates in a reversible reaction that generates 2', 3'-cyclic phosphate and 5'-hydroxyl termini. In constrast to the hammerhead and *Tertrahymena* ribozyme reactions, hairpin-mediated cleavage and ligation proceed through a catalytic mechanism that does not require direct coordination of metal cations to phosphate or water oxygens. The hairpin ribozyme is also a better ligase than it is a nuclease while hammerhead reactions favour cleavage over ligation of bound products (Fedor 2000). Nesbitt et al. (1999) tried to explore the basis for this difference. They found that the hairpin structure is more rigid and undergoes a small decrease in dynamics upon ligation than the more flexible hammerhead structure, thus suggesting that conditions that stabilize RNA structure tend to promote ligation. They further think that hairpin cations promote ligation through de-localized electrostatic shielding, perhaps interacting with a region of especially high charge density in the ligated ribozyme.

Hammerheads

Hammerhead ribozymes represent one class of ribozymes currently developed for therapeutical applications (Fig. 2.21; Symons 1992; Hertel et al. 1996).

To be effective in gene inactivation, the hammerhead ribozyme must cleave a complementary RNA target without deleterious effects from cleaving non-target RNAs that contain mismatches and shorter stretches of complementarity. The specificity of hammerhead cleavage was evaluated using HH16, a well-characterized ribozyme designed to cleave a target of 17 residues. Under standard reaction conditions, HH16 is unable to discriminate between its full-length substrate and 3'-truncated substrates, even when six fewer base pairs are formed between HH16 and the substrate. This striking lack of specificity arises because all the substrates bind to the ribozyme with sufficient affinity so that cleavage occurs before their affinity differences are manifested. In contrast, HH16 does exhibit high specificity towards certain 3' truncated versions of altered substrates that either also contain a single base mismatch or are shortened at the 5' end. In addition, the specificity of HH16 is improved in the presence of p7 nucleocapsid protein from human immunodeficiency virus (HIV-1) which accelerates the association and dissociation of RNA helices. These results support the view that the hammerhead has an intrinsic ability to discriminate against incorrect bases, but emphasizes that the high specificity is observed only in a certain range of helix lengths (Hertel et al. 1996; Figs. 5.2, 5.3)

A variety of drugs inhibit biological key processes by binding to a specific RNA component. Hermann and Westhof (1998) offered a model for the inhibition of the hammerhead ribozyme. In their opinion charged ammonium groups of the aminoglycosides might dock at positions of the hammerhead ribozyme that are occupied by Mg^{2+}. The covalently linked ammonium groups of the aminoglycosides are thus able to complement in space the negative electrostatic potential created by a three-dimensional RNA fold. Murray et al. (1998), were able to capture a conformational change that takes place in the cleavage site of the hammerhead ribozyme during self-cleavage. This rearrangement brings the hammerhead ribozyme from the ground state into a conformation that is poised to form the transition state geometry required for hammerhead RNA self-cleavage

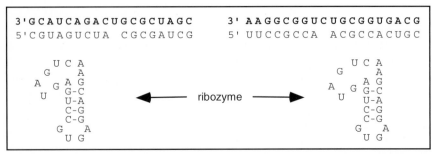

5.2. Targeting of RNA by ribozymes. Two retroviral sequences targeted by hammerhead ribozymes (Sullenger and Cech 1993) are shown *in bold*. In a similar fashion, different viral RNA sequences and motifs are amenable to such a ribozyme-mediated strategy of attack

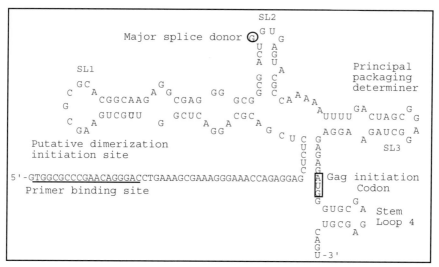

Fig 5.3 This figure showes a secondary structure model of the major HIV-1 packaging signal recognition including stem loops 1, 2, 3, and 4. It consists of an extended bulged stem loop whose structure is altered on interaction with the Gag polyprotein. For further details refer to Zeffman et al. 2000.

.

Ribonucleoprotein Particles (RNPs)

A pseudo-half knot can be formed by binding an oligonucleotide asymmetrically to an RNA hairpin loop, thus altering for example the TAR structure in HIV in such a way that specific recognition and binding of a Tat-derived peptide is disrupted (Ecker et al. 1992). Though this is only a preliminary experiment, the results indicate that specific manipulation of RNPs is a further area where therapeutic modulation of regulatory RNAs may be beneficial, such as in HIV infection (Ecker et al. 1992). The RNP domain that is involved in the recognition of a variety of RNAs has been studied in detail by Mittermaier et al. (1999), who revealed some important aspects of RNP-RNA recognition in human U1A-RNA complexes. Changes in backbone dynamics upon complex formation identify the region of the protein where conformational exchange processes are quenched in the RNA-bound conformations. Furthermore, amino acids whose side-chains experience significant changes in conformational flexibility coincide with residues important for the specificity of the U1A protein / RNA interaction.

RNPs are also a target for several important autoimmune diseases such as Lupus erythomatodes and sclerodermia. Antisera from patients with such diseases have been critical in identifying and cloning several genes responsible, for instance those encoding snRNAs (e.g. Dandekar and Tollevey 1989; and numerous other examples). Unfortunately, the opposite question, how the immune response against RNPs is mounted in the patient, is far less well understood. Animal models for autoimmune RNPs exist, such as the injection of rats with mercury. The study of such models shows that the autoimmune antibodies are probably a side effect of an initially triggered autoimmune reaction; this may also be the case in man. Further details of the reaction including the regulation on the RNA level have unfortunately not been examined, probably due to lack of collaboration between clinically oriented and fundamental research.

Apoptosis

X-linked inhibitor of apoptosis protein (XIAP) is a key regulator of programmed cell death triggered by various apoptotic agents. Translation of XIAP is controlled by a 162 nt internal ribosome entry site (IRES) element located in the 5' UTR of XIAP mRNA. XIAP IRES mediates efficient translation of XIAP under physiological stress and enhances cell protection. Moreover, the La autoantigen is identified as a protein that specifically binds XIAP IRES *in vivo* and *in vitro*. The biological relevance of this interaction is further demonstrated by the inhibition of XIAP IRES-mediated translation in the absence of functional La protein. These points suggest an important role for the La protein in the regulation of XIAP expression, possibly by facilitating ribosome recruitment to XIAP IRES (Holcik and Korneluk 2000).

Suggestions for Further Therapies

To fully exploit the therapeutic potential of RNA motifs and interactions, one first has to examine which diseases are caused by aberrant regulation of RNA or are RNA-mediated. The above examples examined this using stringent criteria, however, in a broader sense, any disease involving proteins is also RNA-mediated and thus in principle amenable to modulation of RNA metabolism.

For many developmental processes, in regulation and differentiation, metabolism of mRNA is important because the messenger is modified during time. These modifications include stability of mRNA and corresponding regulatory elements in the RNA, RNA editing, for instance in messengers for neuronal receptors (see Chap. 6) and the intricate interplay of mRNA localization and translational signals during differentiation and development by different mRNA signals. Each of the regulatory elements involved provides targets for modulation by antisense oligonucleotides and ribozymes as well as antisense therapeutics.

5.8   Perspectives

New Targets

Antisense modulation of mRNA activity is generally thought of as a new therapeutic concept; however, the full extent of antisense regulation present already *in vivo* is probably currently underestimated. The possibilities for finding short antisense stretches of different RNAs which interact with each others' expression are numerous, but it is simply too tedious to check the many potential combinations. Known examples thus concentrate on genes, which are tightly connected and thus an antisense repression can be revealed. In phages, examples of antisense regulation are numerous, e.g. in phage P22 *sar* RNA supresses antirepressor protein (Wu et al. 1987). In bacteria several examples for antisense RNA regulation are also known, e.g. *micF* and *dicF*. In addition, in *C. elegans* there is *lin-4* (Delihas 1995). However, the number of examples in humans is not at all clear. A complete search of all against all, e.g. for complementary 20 base pairs, even allowing for up to three mismatches, is already quite tough for 2000 base pairs, let alone any experimental tests. This however by no means indicates that also antisense inhibtion may occur in not yet discovered instances. Also in other domains where RNA was selected as the favorite molecule to carry out the process, in particular catalysis and information processing, important therapeutic targets have been identified. Clinical applications also include the various illnesses involving mutations in splicing (Lamond 1995), for instance different beta-thalassemias, and of course the disastrous effect of different RNA and DNA viruses. Several antisense inhibition scenarios for HIV have already been described (Sullenger and Cech 1993) but the stumbling block is still efficient delivery *in vivo*. Thus *trans*-splicing in trypanosomes could be such a target for efficient therapeutic and antiparasitic drugs, provided they avoid blocking similar reactions in vertebrates (Dandekar and Sibbald 1990; Bruzik and Maniatis 1992). The different RNA processing steps and the modified RNAs they yield lead to further targets for medical intervention. RNA editing in the Wilms' tumour susceptibility gene has been described (Sharma et al. 1994). Uracil in position 839 of the mRNA is edited to cytosine. However, this isoform of the mRNA is only found in the adult rat kidney in the same amount as the unmodified mRNA whereas it is absent, in the neonatal kidney. Modulation of high specific mRNA editing processes in such genes raises interesting possibilities for novel therapeutic interventions in cancer and differentiation.

Molecules that Bind RNA with Sequence Selectivity

Drug-like molecules that bind RNA with sequence selectivity would provide valuable tools to elucidate gene expression pathways and new avenues to the treatment of degenerative and chronic conditions. Varani et al. (2000), report the investigation of the structural basis for recognition of an RNA stem-loop by neomycin, a naturally occurring aminoglycoside antibiotic. Neomycin binds the RNA stem-loop that regulates alternative splicing of exon 10 within the gene coding for human tau protein. Mutations within this splicing regulatory element destabilize the RNA structure and cause frontotemopral dementia and Parkinsonism linked to chromosome 17 (FTDP-17), an autosomal dominant condition leading to neurodegeneration and death. The three-dimensional structure of the RNA-neomycin complex shows interaction of the drug in the major groove of the short RNA duplex, where familial mutations cluster.

Translational Inhibition

Work on structural RNA in translation such as Olsthoorn et al. (1995), Thanaraj and Argos (1996) and related research (Brunak and Engelbrecht 1996) show that RNA structure is far more important in translation than previously anticipated. However, inhibition of RNA structures involved in translation may also be achieved by the use of binding proteins (e.g. IRP on IREs) and this alley for therapeutic approaches on unwanted RNA molecules via binding proteins is a further potential therapeutic option.

Applying RNA Tools

c-myc expression can be efficiently blocked by peptides (Choo et al. reviewed in Dandekar and Argos 1995). Such peptide blocking of different regulatory RNA structures is another alternative to antisense approaches. Short RNA motifs in viridae, in snoRNAs or hammerhead motifs are reasonable well understood. However, far more medical implications have the RNA structures involved in embryogenesis and oncology; we are only starting to understand these. More research is required, it is especially necessary to elucidate the more general mechanisms. They can function via the 5' and/or 3' UTR and/or suitable stem-loop structures recognized by protein factors. Various different splicing mechanisms can also be envisaged. These have to be differentiated from alternative and more rare levels of regulation such as antisense RNAs acting in *trans* or short ORFs modulating ribosome scanning.

Regulation of transcription in RNA-viruses can be regulated RNA-mediated, although this is unusual and normally transcription is regulated by proteins. An example exists in the red clover necrotic mosaic virus genome. This genome is composed of two single-stranded RNA components, RNA-1 and RNA-2. The viral capsid protein is translated from a subgenomic RNA (sgRNA) that is transcribed from genomic RNA-1 (Sit et al. 1998). RNA-2 is required for transcription of sgRNA. Mutations that prevent base-pairing between RNA-1 subgenomic promoter and the RNA-2 *trans*-activator prevent expression of a reporter gene. So it is thought that binding of RNA-2 to RNA-1 *trans*-activates sgRNA synthesis (Sit et al. 1998).

6. Areas of Research on RNA Motifs and Regulatory Elements

6.1 Introduction

The following subchapters present several areas in which we consider the detection of new regulatory RNAs and, concomitantly with these, new regulatory RNA motifs, to be making impressive progress. Using the techniques and approaches discussed in the previous chapters, the action and interactions of regulatory elements are revealed to be at the heart of the RNA function investigated. Of course, the selection of research areas is always subjective and we apologize for other focus points not mentioned due to limitations of space and time.

A central area of research in the RNA field is the maturation pathway of pre-mRNA (Fig. 1.1). It is first spliced, sometimes with editing, then capped, and polyadenylated. Next it travels through nuclear pores into the cytoplasm where its translation can be regulated via motifs in the 5' UTR, ORF and 3' UTR. This whole process is rich in regulatory RNA elements and motifs and new ones are constantly being discovered. Thus, this RNA pathway, starting from the different forms of splicing, is highlighted first, and subsequently some other areas of current research.

6.2 The Transcription of DNA

A first step of regulation can be located at the level of transcription. Yarnell and Roberts (1999) showed that transcription elongation by RNA polymerase is influenced by regulatory elements. Terminators can disrupt the elongation complex and release RNA, and antiterminators can overcome termination signals. A complementary oligonucleotide that replaces the upstream half of the RNA hairpin stem of intrinsic terminator transcripts induces RNA release in *E. coli*, implying that RNA hairpins act by extracting RNA from the transcription complex. A transcription antiterminator inhibits this activity of oligonucleotides and therefore protects the elongation complex from destabilizing attacks on the emerging transcript.

6.3 The Splicing of Messenger RNA

After transcription by an RNA polymerase, the nascent RNA transcript is processed in the nucleus to yield the mature mRNA, which is exported, and the introns, which remain in the nucleus. The RNA structures involved and RNA interactions with different proteins and other RNAs by virtue of different motifs are a focus of current research. Splicing factors such as U1 recognize their cognate protein by specific loop structures and nucleotides.

The nuclear organization of splicing factors is also interesting, as there are foci in the nucleus where different species are concentrated. SnRNAs, but also components of the splicing machinery, are concentrated in different regions of the nucleus, which partly overlap. Additional studies show that compartmentation is less strict if also lower concentrations of splicing factors are measured (Neugebauer and Roth 1997).

Splicing occurs in eukaryotes and in some kind in archaebacteria (Lykke-Andersen et al. 1994; Belfort 1995). Differential splicing such as in sex-lethal (Kelley and Kuroda 1995) is important for tissue- or species-specific differentiation, it requires recognition of specific regulatory motifs.

A juxtaposed splicing enhancer, and inhibitor has been identified by Kan and Green (1999). This splicing enhancer and inhibitor is located in the exons M1 and M2 of pre-mRNA of IgM. The primary function of the enhancer is to counteract the inhibitor.

Different forms of splicing and RNA recognition

Eukaryotic genes are interrupted by intervening sequences (introns) which are removed, "spliced-out", posttranscriptionally. Introns are cut out in a two step transesterification reaction (Ruskin et al. 1984; Konarska et al. 1985; Moore and Sharp 1993; Fig. 2.10). The first step involves cleavage at the 5' splice site, in which the 2' OH-group of an internal "branch-point" adenosine residue attacks the 5' phosphate residue at the splice site. This results in formation of a lariat-intermediate and a free 5' exon. The second step involves a nucleophilic attack of the 5' exon on the 3' splice site. This leads to the formation of the ligated exons and the release of the intron as a lariat-intermediate (for reviews, see Madhani and Guthrie 1994; Burge et al. 1999; Mistelli 2000). These reactions take place in a complex ribonucleoprotein particle, termed the spliceosome, and require several RNA/RNA and RNA/protein interactions. In metazoa, assembly of the spliceosomal U snRNPs requires nuclear export of U snRNA precursors. This export is dependent on the RNA cap structure, nuclear cap-binding complex (CBC), the export receptor CRM1/Xpo1, and RanGTP as well as the lately discovered PHAX (phosphorylated adapter for RNA export). PHAX is phosphorylated in the nucleus and after exporting it becomes dephosphorylated, causing disassembly of the export complex (Ohno et al. 2000). Key players in the spliceosome are the five small nuclear RNAs U1, U2, U4, U5 and U6, which are known to be essential (Moore et al. 1993; Newman 1994). For splicing to occur with high fidelity, the splice sites and the branch point must be recognized. It has long been known that the U1 snRNP base pairs at the 5' splice site at a conserved intronic sequence, the canonical GUAUGU (see Chap. 2) (Lerner et al. 1980; Rogers and Wall 1980). Interestingly, *Schizosaccharomyces pombe* U1 snRNP has in addition been shown to bind concomitantly to conserved nucleotides at the 3' splice site (Reich et al. 1992). This might then lead to the juxtaposition of both the 3' and 5' splice sites for splicing to occur in a concerted fashion (Steitz 1992). U2 snRNP interacts with the branch point region (Black et al. 1985), in which the adenosine forms a bulge in the U2/intron duplex (Parker et al. 1987). U2 binding to the branch point requires the involvement of several proteins, which are needed to stabilize a specific stem-loop structure (stem II A) within the U2 RNA, thus allowing branch point recognition (Ares and Igel 1990; Zavanelli and Ares 1991; Figs. 2.11, 2.12). The splicing factor U2AF is made of the large U2AF65 and the small U2AF35 subunit. It faciliates the interaction of U2 with the splicing branch point. Recent analyses from Wu et al. (1999) and Zorio and Blumenthal (1999) showed that U2AF65 binds to the poly(Y) tract while U2AF35 contacts the 3' splice site and recognizes the AG/G consensus. Initial branch point recognition (the motif UACUAAC in yeast) is effected by SF1 in yeast or BBP in mammals (the mammalian orthologue, branch point binding or bridging protein) (Berglund et al. 1997). After the binding of U1 and U2 to the pre-mRNA, U4, U5 and U6 bind as a preformed tripartite complex. U4 and U6 base pair with each other (Hashimoto and Steitz 1984; Rinke et al. 1985; Brow and Guthrie 1988; Fig. 2.11). The interaction of the U5 snRNP with the U4/U6 snRNP is still poorly understood. This interaction is, however, abolished prior to the catalytic steps of splicing (Lamond 1985; Yean and Lin 1991), and might be mediated by RNA-dependent helicases. During catalysis, several RNA/RNA interactions are disrupted and reform with new partners, leading to a dynamic change in the overall spliceosome structure (Madhani and Guthrie 1994; Nilsen 1994a). Crosslinking studies have shown that U6 base pairs with conserved intronic sequences at the 5' splice site (Wassarman and Steitz 1992). The previously attached U1 snRNP is displaced (Kandel-Lewis and Seraphin 1993). The U4/U6 base pairing is also disrupted and is replaced by the formation of two alternative interactions. The first is an intermolecular interaction between U2 and U6, the second is an intramolecular interaction within the U6 helix (Madhani and Guthrie 1994). The U6/U2 (helix I) duplex is postulated to comprise the catalytic active site (Fig. 2.11). Such an interaction leads to the direct juxtaposition of the reactants of the first catalytic step. Mutations that interfere with this interaction indicate that it is required for both splicing steps (Fabrizio and Abelson 1990; Madhani and Guthrie 1992; McPheeters and Abelson 1992). U5 snRNP plays a significant role during the second step of catalysis in which the two exons are ligated and the intron is released. A conserved stem loop structure within the U5 snRNA can base pair with the 5' exon and concomitantly also base pair to the 3' exon (Newman and Norman 1992; Sontheimer and Steitz 1993). These simultaneous interactions may hold the two exons in close proximity to each other and enable their ligation (Sontheimer and Steitz 1993).

In archea the crystal structure of a dimeric splicing endonuclease has been reported (Li and Abelson 2000). The structures of the N-terminal and the C-terminal repeats, the binding region for RNA substrates containing a bulge-helix-bulge motif could be determined. Based on the identified RNA-binding region, a cation-pi interaction is suggested to be responsible for coordinating activities between the two active sites.

In metazoan, two distinct spliceosomes catalyzing pre-mRNA splicing are known. The minor spliceosome (U12-dependent) has the human U11/U12 snRNP as a subunit. It has been found that subsets of the spliceosomal proteins are unique to the minor spliceosome or unique to both (Will et al. 1999).

ATAC Introns

A recent interesting finding was the discovery of a minor class of introns, which contain the conserved nucleotides AT, at the 5' end of the intron and AC, at the 3' splice site, in contrast to the GTAG-introns (Fig. 6.1; Hall and Padgett 1994; Mount 1996; Kreivi and Lamond 1996).

So far only a few metazoan genes have been identified which contain these ATAC-introns (called "attack" introns, though, of course, the gene is translated as AUAC in the pre-mRNA), which include P120, CMP, REP-3, CDK-5, SCN4A, SCN5A and *prospero* (Wu and Krainer 1996). Splicing of these introns requires the corresponding rare class of snRNAs U11 and U12 (Hall and Padgett 1996; Tarn and Steitz 1996a). U12 has been shown to base pair with the branch point of these ATAC-introns (Hall and Padgett 1996). Mutations in the consensus branch point sequence (UCCUUAAC in the RNA) prevent splicing *in vivo*, whereas compensatory base pair changes in U12 restore splicing, indicating that the U12/branch point interaction is absolutely required for splicing. This occurs in a manner analogous to the U2/branch point interaction in the major class of introns. Using an *in vitro* system in which they were able to reproduce splicing events of the ATAC-intron from the human P120 gene Tarn and Steitz (1996a) obtained similar results. Upon depletion of the U2 snRNA from extracts using antisense 2' O-methyl oligonucleotides, splicing of the P120 pre-mRNA was activated.

Employing psoralen crosslinking, they established that the U12 RNA binds to the branch point region. Analysis of the spliceosome content further established the involvement of U11, U12 and U5 snRNA, but not U4 and U6. Thus, it was thought that U5 was a common component shared by both classes of spliceosomes.

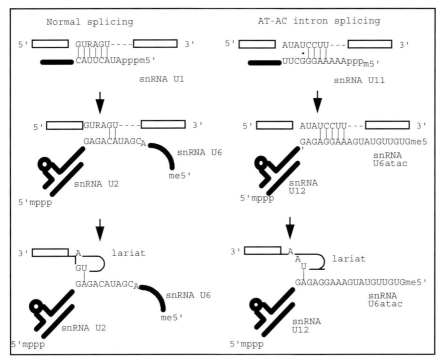

Fig. 6.1. Comparison between normal splicing and splicing of the minor class of ATAC introns. U1,U2 and U6 interactions with the pre-mRNA are shown schematically. Also drawn are the postulated homologous interactions in AT-AC introns, showing important nucleotides and secondary structure features in the newly discovered snRNAs U11, U12 and U6atac. (Yu and Steitz 1997).U6-U2 and the homologous U6atac-U12 interactions are drawn schematically

This was quite surprising as U6 was thought to be involved in the catalytic process (Fabrizio and Abelson 1990; Madhani et al. 1990). Recently, however, Tarn and Steitz (1996b) have identified two novel human snRNAs (termed U4atac and U6atac) which closely resemble the U4 and U6 of the major class of spliceosomes. These two snRNAs are able to recapitulate all the RNA/RNA interactions which are seen in the GUAG spliceosome (see also Nilsen 1996). Base pairing between U12 and U6atac occurs, reminiscent of the U6/U2 helix I, and is important for the catalysis function (Fig. 6.1). It has now also been shown that the 5' splice site of the ATAC-introns is recognized by U11 and U6atac, analogous to the U1 and U6 interactions in splicing of the major class of introns (Yu and Steitz 1997). Recently it has been shown that U1 is involved in the splicing of an ATAC sodium channel intron (SCNA4) (Wu and Krainer 1996). This indicates that the two different types of spliceosomes may even act in concert during splicing.

Another point of interest is the tRNA splicing. The specificity for the recognition of the tRNA precursor resides in the endonuclease, which removes the intron by making two independent endonucleolytic cleavages. Although the enzymes in eukaryotes and archea appear to use different features of pre-tRNA to determine the site of cleavage, Fabbri et al. (1998) showed that the eukaryotic enzyme is able to use archeal recognition signals. This indicates that a common ancestral mechanism for recognition of pre-tRNA by proteins might exist.

Handa N et al. (1999) offer structure insights as to how a protein binds specifically to a cognate RNA without any intramolecular base pairing. They studied the sex-lethal protein (Sxl) of *D. melanogaster* that regulates alternative splicing of the transformer (tra) mRNA precursor by binding to the tra polypyrimidine tract during the sex-determination process. The crystal structure revealed that the RNA is characteristically extended and bound in a beta-sheet cleft, where the UGUUUUUUU sequence is specifically recognized by the protein. Penalva et al. (2000) report that *Drosophila* fl(2) gene is required for female-specific splicing of Sxl and tra pre-mRNAs. The fl(2) gene encodes for a nuclear protein with an HQ-rich domain. Puoti and Kimble (2000) revealed by phylogenetic analysis that the *C. elegans* MOG-1, MOG-4 and MOG-5 proteins are closely related to the yeast proteins PRP16, PRP2 and PRP22, respectively. They propose that MOG-1, MOG-4, and MOG-5 are required for post-transcriptional regulation, perhaps by modifying the conformation of ribonucleoprotein complexes.

6.4 *Trans*-Splicing

Another important area is *trans*-splicing. This was originally considered a rare side reaction occurring in certain parasites such as trypanosomes. However, genomic searches were able not only to identify the known examples of *trans*-splicing but also to indicate a large potential for these and related reactions to occur in other organisms including vertebrates (Dandekar and Sibbald 1990). These results were followed up by Bruzik and Maniatis (1992), who showed that the leader sequence from parasites could, in fact, be *trans*-spliced by factors contained in HeLa cell extracts and a number of examples from a wide range of different organisms (Nilsen 1995; Bohnen 1993)

Trans-splicing involves the RNA splicing of separate precursors giving rise to one mature mRNA. One can distinguish two different forms of *trans*-splicing: 1) the discontinuous group II intron form and 2) the spliced leader form of *trans*-splicing (Bohnen 1993). The discontinous group-II intron form of *trans*-splicing is generally found in plant and algal chloroplasts and mitochondria (Goldschmidt-Clermont et al. 1991; Chapdelaine and Bohnen 1991). Splicing brings two different RNA precursors together and is thought to be dependent on the recognition of the group-II intronic structures of the precursor transcript, akin to nuclear pre-mRNA splicing (Jacquier 1990; Jacquier and Jacquesson-Breuleux 1991). The spliced leader type of *trans*-splicing is found in protozoans, such as trypansomes, and in nemadotes, in which short 5' capped non-coding sequences are attached to the mRNA. It is quite similar to nuclear pre-mRNA splicing and occurs in a very similar RNP complex (Guthrie 1991). In addition to U-snRNPs, however, a spliced leader (SL)-RNA (Sharp 1987) is required. This contains an intronic portion, which may provide part of a functional equivalent to the U1 snRNA in nuclear pre-mRNA splicing (Bruzik and Steitz 1990). But that may be better provided by still to be discovered novel snRNAs.

C. elegans genes are *trans*-spliced by receiving a 22 nt SL sequence at their 5' ends (Krause and Hirsch 1987). The leader is derived from a 100 nt RNA which itself is part of an snRNP with a functional Sm site (see Chaps. 2, 4) as in other snRNAs (Thomas et al. 1988; Bektesh et al. 1988; Ferguson et al. 1996). Unlike in trypanosomes, *cis*- and *trans*-splicing both occur in *C. elegans*. It has been shown that if an mRNA transcript begins with an intron like RNA at its 5' end (termed outron), rather than an exon then it is targeted for *trans*-splicing (Conrad et al. 1991). Interestingly *C.elegans* has developed two SL forms, i.e. SL1 and SL2, in which the SL1 is *trans*-spliced onto most of the mRNA, but SL2 from snRNA SL2 can provide the same function if SL1 is missing (Bektesh et al. 1988; Kuwabara et al. 1992; Ferguson et al. 1996).

What would be the physiological significance of *trans*-splicing in higher eukaryotes, for instance in man? An interesting indication is another RNA modification process, RNA editing (Fig. 6.2). Originally, this was also only discovered in parasites. However, subsequently, interesting vertebrate examples were identified, such as the glutamate receptor RNA (Fig. 6.2). Considering this example and also the estimate that about half of the roughly 35000-45000 human genes are involved in brain differentiation, the *trans*-splicing type of RNA regulation may also be implicated in similar processes, including development and differentiation. Splicing enhancers can control splice-site selection and alternative splicing of nuclear pre-mRNA. Bruzik and Maniatis (1995) showed that RNA molecules containing a 3' splice site and enhancer sequence are efficiently spliced in *trans*. The products are RNA molecules containing normally *cis*-spliced 5' splice sites or normally *trans*-spliced SL-RNAs from lower eukaryotes. Ser- Arg-rich splicing factors bind and stimulated this reaction.

Recent research indicates more directly that *trans*-splicing processes occur in mammals: Shimizu (1996) describes *trans*-splicing as the most probable mechanism involved in second isotype immunoglobulin expression simultaneously expressed with IgM. Further, *trans*-splicing and alternative-tandem- *cis*-splicing are two ways by which mammalian cells generate a truncated SV40 T-antigen (Eul et al. 1996). *Trans*-splicing has also been shown to occur in cultured mammalian cells (Ferguson et al. 1996).

Clearly identified *trans*-splicing reactions include *trans*-splicing with SL sequences (Bohnen 1993), such as:

- In trypanosomes (all mRNAs; 39 nt mini-exon within 140 nt SL-RNA; SL-RNA genes in tandem arrays).
- In nematodes (10-15% of mRNAs; 22 nt mini-exon within 100 nt SL-RNA; SL1 (but not SL2) RNA genes in tandem repeats with 5S rRNA genes, in some genera on opposite strands).

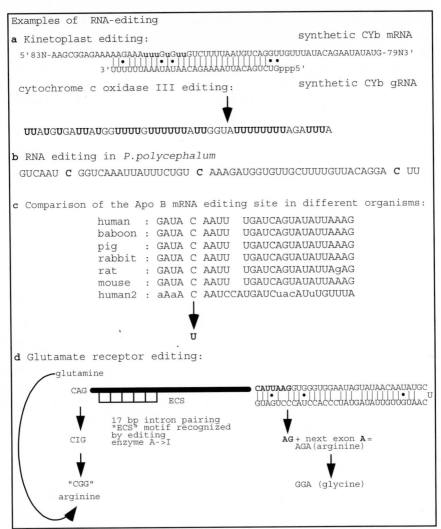

Examples of RNA-editing

a Kinetoplast editing: synthetic CYb mRNA

5'83N-AAGCGGAGAAAAAGAAA**uuuGuGuu**GUCUUUUAAUGUCAGGUUGUUUAUACAGAAUAUAUG-79N3'

3'UUUUUUAAAUAUAACAGAAAAUUACAGUCUGppp5'

cytochrome c oxidase III editing: synthetic CYb gRNA

UUAUGUGAUUAUGGUUUUGUUUUUUAUUGGUAUUUUUUUUAGAUUUA

b RNA editing in *P.polycephalum*

GUCAAU **C** GGUCAAAUUAUUUCUGU **C** AAAGAUGGUGUUGCUUUUGUUACAGGA **C** UU

c Comparison of the Apo B mRNA editing site in different organisms:

```
human  : GAUA C AAUU  UGAUCAGUAUAUUAAAG
baboon : GAUA C AAUU  UGAUCAGUAUAUUAAAG
pig    : GAUA C AAUU  UGAUCAGUAUAUUAAAG
rabbit : GAUA C AAUU  UGAUCAGUAUAUUAAAG
rat    : GAUA C AAUU  UGAUCAGUAUAUUAgAG
mouse  : GAUA C AAUU  UGAUCAGUAUAUUAAAG
human2 : aAaA C AAUCCAUGAUCuacAUuUGUUUA
```

U

d Glutamate receptor editing:

glutamine

CAG   **CAUUAAG**GUGGGUGGAAUAGUAUAACAAUAUGC
            ECS   GUAGUCCCAUCCACCCUAUGAUAUUGUUGUAAC U

17 bp intron pairing
"ECS" motif recognized
by editing
enzyme A->I

CIG   AG + next exon **A** =
                                           AGA(arginine)

"CGG"   GGA (glycine)

arginine

Fig. 6.2a-d. Examples of RNA editing. **a** Kinetoplast editing in *Trypansoma brucei*. A prerequisite for kinetoplastid mRNA editing is the formation of a heteroduplex between the mRNA and a guide RNA, as has been established in vitro by Harris and Hadjuk (1992).Additional uracils, shown *in bold*, are incorporated into the mRNA during editing. Editing of cytochrome c oxidase III leads to the insertions of several uracils (*bold*) and the deletion of a single uracil (indicated by an *arrow*). **b** Editing in Physarum polycephalum. Inserted cytosines are indicated *in bold* (Bass 1991). **c** Editing of the Apo B mRNA site (C into U). Several mammalian mRNA editing sites and a second human editing site are compared. A close consensus becomes apparent. **d** Glutamate receptor editing. A double-stranded structure in the receptor-mRNA mediates an A to G editing at the Q/R-site by enzymatic conversion of adenine to inosine which changes a glutamine codon to arginine during translation. Similarly, a second site termed the R/G site is edited, mediating an A to G change by a secondary structure formed between exon 13 and the intron (exon nucleotides shown *in bold*). This changes an arginine codon into a glycine codon.The significance of these changes is discussed in Chapter 6

- *Schistosoma mansoni* (some mRNAs only are *trans*-spliced using a 36 nt min-exon within 90 nt SL-RNA; SL-RNA genes again in tandem arrays).
- *Euglena gracilis* where most mRNAs are *trans*-spliced using a 26 nt mini-exon within 100 nt SL-RNA (SL-RNA genes in tandem repeats with 5S rRNA genes).
Examples for discontinuous group II intron splicing are:
- The chloroplast *rps12* gene in land plants (liverwort, tobacco) with one *trans*- and one *cis*-spliced intron.
-In *Chlamydomonas* chloroplasts the *psaA* gene, which has two *trans*-spliced introns; tcsA RNA is required for splicing of the first discontinuous intron.
-The *nad* genes 1-3 in mitochondria of higher plants, where two or three out of four introns in each gene are *trans*-spliced.

6.5   RNA Editing

The central dogma of biology states, that genetic information flows from DNA to RNA in a continuous manner, without incorporating any changes. This holds true only in part, as has been shown by the discovery of the RNA-editing phenomenon. RNA editing involves deletion, insertions or substitutions of nucleotides in RNA templates, thereby altering the expression of genetic information. In a biological context, editing can have profound effects on downstream events, as it might generate proteins with different properties or alter non-coding regions so that the message becomes translatable. Examples of editing have been found in various different organisms ranging from trypanosomes (Stuart 1991, 1993; Hajduk et al. 1993) to mammals (Hodges and Scott 1992; Smith 1993). In flowering plants (Covello and Gray 1989; Gualberto et al. 1989; Kudla et al. 1992; Maier et al. 1992) and slime moulds (Bass 1991; Mahendran et al. 1991; Miller et al. 1993). RNA editing in viral RNAs has been described by Luo et al. 1990; Casey et al. 1992; Zheng et al. 1992. Besides delta-virus there is also editing in measles virus (Vanchiere et al. 1995). Most examples of RNA editing described occur in the mitochondria; however, editing of transcripts, which encode nuclear genes has also been demonstrated. Mammalian mRNAs, which are posttranscriptionally modified via editing, are those of apolipoprotein B and of glutamate receptor channels, mediating rapid excitatory neurotransmisson in the brain (Fig. 6.2).

Mammalian RNA Editing

Apolipoprotein B (apo B) transports cholesterol and triglycerides in the blood (Scott et al. 1989) and is present in two different forms i.e. apo B100 and apo B48 (Chen et al. 1987; Powell et al. 1987). By editing of apo B mRNA the small isoform apo B48 is generated. Editing of apo B not only varies considerably between different tissues (Boström et al. 1990), but is also affected by hormones and nutrition (Davidson et al. 1988; Baum et al. 1990). Two editing sites are known in the mRNA, in which cytosines at positions 6802 and 6666 are modified to uracil by deamination (Chen et al. 1987; Powell et al. 1987; Nawaratnan et al. 1991). This change converts a CAA glutamine to a UAA stop codon, producing a variant of the protein with shorter length, leading to quicker delivery of dietary lipids to the liver (Brown and Goldstein 1987; Fig. 6.2). A stretch of 11 nucleotides (5'-UGAUCAGUAUA-3') downstream of the edited nucleotide is an important determinant, as mutations in this region either abolish or greatly diminish the editing capacity *in vitro* (Shah et al. 1999). APOBEC1 is the liver-specific enzyme, which is responsible for editing of the C to U in the apolipoprotein B mRNA. Overexpression of this enzyme has been shown to cause liver cancer in transgenic mice. It was thought that this was the result of erroneous editing of one or more mRNAs. Recently, Yamanaka et al. (1997) have identified such a target transcript, termed NAT1 for novel APOBEC target1. The message codes for a protein, which is similar to the translation factor eIF4G. These messages which usually do not undergo editing, are hyperedited and accumulate termination codons in transgenic mice developing liver cancer.

Glutamate-gated ion channel subunits or glutamate receptors (GlnR) also undergo a post-transcriptional modification in which adenosine residues are converted to inosine in the mRNA by a single-step deamination (Kim and Nishikura 1993; Bass 1995). Inosines are read as guanosines by the translation machinery (Basillo et al. 1962). Thus, this editing alters the gene-encoded glutamine (Q) codon CAG to CGG, which now encodes for an arginine (R). Editing at this Q/R site, requires formation of a 17 base pair double stranded RNA (dsRNA) structure between exon sequences and an exon-complementary intronic sequence (ECS, Fig. 6.2), located several hundred nucleotides downstream in the adjacent intron (Higuchi et al. 1993; Egebjerg et al. 1994; Herb et al. 1996). The result of this editing is a dramatic decrease in Ca^{2+} permeability of these excitatory receptors. Lomeli et al. (1994) identified an additional site within the receptor GluRB, which is also edited. This has been termed the R/G site, in which a codon switch from AGA (arginine) to GGA (glycine) occurs, leading to faster recovery kinetics from channel desensitization. Thus, RNA editing here is regulated by specific RNA motifs and serves to create receptors with different properties from the same RNA transcript, illustrating that the process of RNA editing can create plasticity similar to that generated by splicing. These and other RNA processes are probably also involved in generating receptor diversity in the nervous system (more than 10^{10} neurons in the human brain), while only a limited set of genes can successfully be passed to the next generation with the existing accuracy of the mammalian replication apparatus.

Another interesting example concerning adenosine deamination, RNA editing and alternative splicing is present in rat. The enzyme ADAR2 is a double-stranded RNA-specific adenosine deaminase. Several rat ADAR2 mRNAs are produced as the result of two distinct alternative splicing events. One of them uses a proximal 3' acceptor site, adding 47 nts to the ADAR2 coding region. Proximal and distal of this alternative 3' acceptor site adenosine-adenosine (AA) and adenosine-guanosine (AG) dinucleotides are localized. The use of this acceptor site depends upon the ability of ADAR2 to edit its own pre-mRNA by converting intronic AA to AI to mimic the highly conserved AG sequence found at 3' splice junctions. This can be seen as a novel regulatory RNA strategy by which a protein can modulate its own expression (Rueter et al.1999).

Editing in Trypanosomes

RNA editing in *T. brucei* leads to deletion and addition of uridine residues in many of the mitochondrial RNAs (Hajduk et al. 1993; Stuart 1993; Fig. 6.2). The number of edited residues varies largely. It ranges from insertion of only four uridines in the cytochrome C oxidase subunit II RNA (Simpson and Shaw 1989; Benne 1990; Stuart 1991) to large uridine insertions and deletions of a number of RNAs (Bhat et al. 1990; Koslowsky et al. 1990). How is this editing carried out mechanistically? It has been shown that small transcripts function as guide RNAs (gRNAs). These are 55 to 70 nts in length and mediate the addition of uridines by base pairing to the edited sequences (Sturm and Simpson 1990; Blum et al. 1991). gRNAs contain sequences at their 5' end that base pair to mRNA sequences 3' of an editing site, followed by internal sequences and a non-coded 3' poly(U) tail. Blum et al. (1991), proposed that the mechanisms of insertion and deletion is based on a series of transesterification reactions, related to RNA splicing reactions. In the first step the 3' OH group of the 5' gRNA (poly U) attacks the phosphodiester bond at an editing site forming a gRNA/mRNA duplex and a 5' fragment of the mRNA. In the second step the 3' OH group of the 5' mRNA attacks the phosphodiester bond of uridine in the gRNA. This leads to insertion of a uridine in the mRNA and reduction in the length of the gRNA by one nucleotide (Bass 1991). Editing is developmentally regulated during the *trypanosoma* life cycle in which certain forms of edited mRNAs will be present in the insect form and not in the bloodstream form and vice-versa (Feagin et al. 1987; Feagin and Stuart 1988; Koslowsky et al. 1990). This allows the parasite to regulate the production of components of the respiratory system, which it requires during the specific stages of its life cycle.

Editing in the Slime Mould Physarum

RNAs of the slime mould *Physarum polycephalum* show a different kind of editing in which cytidine residues are inserted. These insertions create ORFs in mRNAs and are required for functional structure formation in rRNAs and tRNA (Miller et al. 1993). The alpha-subunit of ATP synthase is edited at 54 sites by insertion of cytidines (Mahendran et al. 1991). Interestingly, the editing sites are evenly distributed in the coding region with an average spacing of 26 nts. Most of the editing sites were shown to be 3' to purine-pyrimidine dinucleotides and at a third position of the codon (Bass 1991). It is possible that editing may again be regulated for adaptation during the life cycle.

RNA Editing in Plant Mitochondria and Chloroplasts

Extensive editing also occurs in plant mitochondria and chloroplasts (Gray and Covello 1993; Schuster et al. 1993). As in *Physarum* and mitochondria of *trypansomes*, plant mitochondria and chloroplast mRNAs are edited by C to U and U to C transitions (Hoch et al. 1991; Kudla et al. 1992). RNA editing is required in certain cases for the generation of functional proteins, e.g. the Cu^{2+}-binding site of the cytochrome oxidase subunit II (cox II) in wheat mitochondria can only be synthesized after editing of its transcript (Covelo and Gray 1990; Gray et al. 1992). Group-II introns are also found in mitochondria, these are also edited to allow proper folding of the RNA to take place to ensure structural integrity, similar to the editing observed in tRNA and rRNAs of trypanosomes. Editing sites are clustered in conserved domains, which, in turn are thought to enhance base pairing interactions (Knoop et al. 1991; Wissinger et al. 1991; Binder et al. 1992). Interestingly, editing in plant mitochondrial mRNAs provides insights into how organelle to nucleus gene transfer took place during the endosymbiotic evolution. It is known that the cox II gene has been transferred from the mitochondria to the nucleus in legumes (Nugent and Palmer 1991; Covello and Gray 1992). If the gene transfer occurred via DNA, then it should be similar to the unedited mitochondrial DNA, but if transfer occurred via an RNA intermediate, one would expect it to resemble the edited mRNA. Indeed, edited sites are found in the nuclear DNA, indicating that the flow of information occurred via an RNA intermediate (Nugent and Palmer 1991; Covello and Gray 1992). To date, the search for gRNAs akin to those found in trypanosomes has been unsuccessful in plant mitochondria and chloroplasts. It is unclear what the biological advantage of this editing process might be.

Editing in Hepatitis Delta Virus

Hepatitis delta virus (HDV) is a subviral human pathogen which requires the concomitant infection of the hepatitis B virus (HBV) to provide coat proteins for packaging (Rizetto et al. 1980a, 1980b). The HDV protein product, also called delta antigen, is synthesized in two forms, depending on the editing event in the amber/w site, which converts a stop codon UAG into a tryptophan codon UGG leading to proteins of 24 and 27 kDa, respectively (Bergmann et al. 1986; Bonino et al. 1986). It was found that the large delta antigen inhibits replication and is required for packaging (Chao et al. 1990; Chang et al. 1991; Glenn and White 1991), whereas the small antigen is required for HDV genome replication (Kuo et al. 1989). It is thought that the large delta antigen curbs replication, thereby allowing the host to survive as a vector for propagation of the disease.

Adenosines are modified to inosines by the ubiquitous nuclear double-stranded RNA-adenosine-deaminase (dsRAD) also called DRADA, which has high affinity for small double stranded ECS sequences (Kim and Nishikura, 1993; Bass 1995; Rueter et al. 1995; Dabiri et al. 1996). It has recently been reported that a purified

DRADA can edit HDV antigenomic RNA *in vitro* (Benne 1996; Polson et al. 1996). DRADA is capable of editing the amber/w site and the R/G site of GluRB mRNA (Dabiri et al. 1996; Melcher et al. 1996; Polson et al. 1996). Hot spots of editing have also been observed, with the 5' nearest neighbor, preference of DRADA being -A=U>C>G- (Polson and Bass 1994).

Ribosomal RNA Processing and Ribosome Biogenesis

A Focus of Interest
RNA processing pathways other than mRNA pathways are another focus of research. In particular, ribosomal processing (Fig. 6.3) is more amenable to detailed studies than before, since more sophisticated techniques to characterize the different RNA species are now available. Twenty years ago, when ribosomal processing was first discovered, the focus was mainly on the proteins involved. However, many RNA structures are involved in the specific processing steps and we will centre on these in the following. This includes a variety of different RNAs and specific, sometimes complex and/or not yet characterized rRNA motif direct cleavages. Several small RNAs are typically located in the nucleolus and therefore called snoRNAs. Important motifs involved in their structure have recently been characterized which now enables rapid detection of new snoRNAs in genomic sequences (Chap. 4). Several of them are intron encoded (Bachellerie et al. 1995; Kiss-Laszló et al. 1996). The discovery of new snoRNAs has been further speeded up by the availability of the complete sequence for the yeast genome. Figure 2.15 shows an overview on snoRNAs and different classes. Several snoRNAs are involved in ribosomal processing and cleavage of rRNA. Further there are methylation guide snoRNAs (Nicoloso et al. 1996) and pseudouridylation guide snoRNAs (Kiss-Laszló et al. 1996; Ganot et al. 1997).
Figure 2.16 summarizes typical base pair interactions between rRNA and methylation guide snoRNAs.
rRNA synthesis occurs in the nucleolus; several snoRNAs are involved in the selection of the different cleavage sites of rRNA precursor during processing to finally yield the mature rRNA species. Ribosome processing and tRNA processing are connected in prokaryotes, the ribonucleoprotein particle RNAse P cleaves out a tRNA situated in the long ribosomal precursor transcript and this aids in the maturation of the rRNA. According to a recent proposal by Morrissey and Tollervey (1995), the eukaryotic ribonucleoprotein particle RNase MRP involved in rRNA maturation may have arisen in eukaryotes as a form of RNAse P specialized on rRNA processing. Instead, a tRNA insert there is simply an internal transcribed sequence present in eukaryotes which is cut out during rRNA maturation. Spacers flanking the bacterial rRNA molecules form extensive helices, which contain the recognition sites for RNase III. Morrisey and Tollervey (1995) further propose that in eukaryotes snoRNPs (such as U3, U14, snR30 in yeast and U8 in *Xenopus*) have replaced intramolecular base pairing in bringing the 5'- and 3'-flanking sequences of the mature rRNAs together.

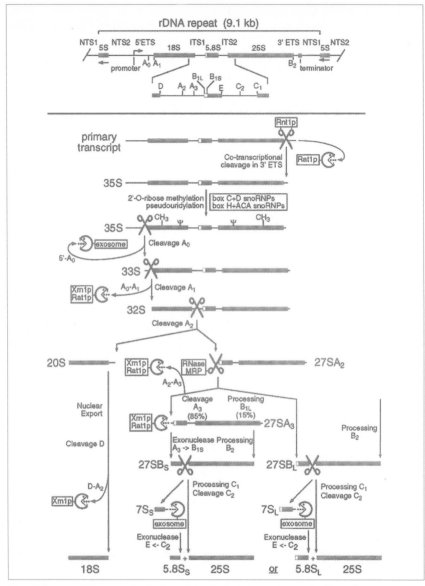

Fig. 6.3 Pre m-rRNA processing in S. cerevisiae. Upper part: The yeast rDNA consists of the 35S pre-rRNA operon and two nontranscribed spacers (NTS1 and NTS2), interrupted by the 5S rRNA gene. The sequences for the mature 18S, 5,8S, and 25S rRNAs, separated by two internal transcribed spacers (ITS1 and ITS2), and flanced by two external transcribed spacers (5'ETS and 3'ETS) are located in the 35S pre-rRNA. The positions of known processing sites are indicated. Lower part: The co-transcriptional cleavage in the 3'ETS generates the 35S pre-rRNA, which becomes modified at numerous sites (schematically depicted as y and CH₃). Processing of the 35S precursor to mature rRNAs occurs by the indicated pathway. Endo- and exonucleolytic activities are represented by "*pacmans*" and *scissors* (After: Venema and Tollervey, 1999)

This would be similar to the independent RNPs interacting at different steps of splicing which evolved in the processing of mRNA to replace the intramolecular interactions found in self-splicing group-II introns.

We as well as others suggested earlier that rRNA is far more active than previously anticipated. A number of results indicate important roles for 16S and 23S rRNA in central ribosomal functions such as tRNA selection and binding, translocation and notably peptidyl transferase activity (Green and Noller 1997). This hypothesis has now been vindicated by the detailed structure of the large ribosomal subunit (Ban et al. 2000). Culver et al. (1999) report that the 7.8 Å crystal structure of the 70S ribosome reveals a discrete double-helical bridge (B4) that projects from the 50S subunit, making contact with the 30S subunit. Moreover, the previously determined structure of a homologous stem-loop from U2 snRNA fits closely to the electron density of the bridge. It will be worthwhile to compare the highly complex structure of 23S rRNA with the ribozyme selected artificially for a peptidyl transferase reaction by Lohse and Szostack (1996) to learn more about required functional features.

The assembly of 80S ribosomes starts on cellular mRNAs with the recruitment of the small ribosomal subunit in the form of a 43S translation pre-initiation complex (Sachs et al. 1997; Gingras et al. 1999). The 5' m^7GpppN-cap structure and its bound initiation factors (eIF)4E, 4G, 4A direct the 43S complex (40S ribosomal subunit, eIF3, eIF2/aaMet-tRNA/GTP) to the 5' end of the mRNA. Subsequently, the pre-initiation complex scans the 5' UTR (involving further eIFs to identify the translation initiation codon to which the 43S complex repositions (Kozak 1989). The final release of initiation factors and the joining of the 60S ribosomal subunit to the 40S one to form a stable, translation-competent 80S ribosome necessitates GTP hydrolysis on both eIF2-GTP and eIF5B-GTP (Merrick and Hershey 1996; Pestova et al. 2000). mRNA translation requires interaction between the anticodon of initiator Met-tRNA, and the cognate start codon of the mRNA. For different cellular mRNAs it is known that initiation of translation can also occur in a cap-independent way on internal ribosome entry sites (IRES) (Jackson 1991; Holcik et al. 2000). Also genomes of many RNA viruses are expressed by IRES-dependent translation initiation. A striking feature of several viral IRESs is their very limited requirement of eIFs to initiate translation (e.g. the IRES of picornaviridae, like poloivirus and encephalomyocarditis virus (EMCV); HCV and the classical swine fever virus (CSFV) (Pestova et al. 1998; Belsham and Jackson 2000; Fig. 2.19). The currently most interesting case is the IRES in the intercistronic region (IGR) of the cricket paralysis virus (CrPV) genome (Fig. 2.19). The CrPV-IGR-IRES drives translation initiation via a recently discovered, unusual mechanism that involves direct 80S ribosome formation without the requirement of any of the eIFs (Wilson et al. 2000a, 2000b). A part of the extended secondary structure of the CrPV-IGR-IRES is thought to mimic a proline tRNA anticodon, which is recognized by the 80S ribosome positioned by the IGR-IRES in its P-site to the corresponding CCU triplet. The first amino acid of the translated peptide-chain is an alanine, encoded by the following GCU triplet in the A-site of the 80S ribosome. aaMet-tRNA, eIF2 and GTP hydrolysis are not required for this type of translation initiation. In mutagenesis experiments it could be shown that the P-site is not decoded and translation is initiated in the A-site of the ribosome (Wilson et al. 2000a, 2000b).   This suggests that the repertoire of translated ORFs in eukaryotic mRNAs may be greater than anticipated (Wilson et al. 2000b).

6.6   Ribosomal RNA Processing Steps

Ribosome biogenesis takes place in a specialized compartment, the nucleolus. Synthesis of rRNA requires a series of post-transcriptional processing steps including   modifications,   endonucleolytic   cleavages   and   exonucleolytic degradation reactions (Eichler and Craig 1994; Klootwijk and Planta 1989). The structure of the rDNA operon is similar in most species in all kingdoms, exceptions   not   withstanding   such   as   the   organization   of   the   rDNA   in archaebacteria. 35S rRNA is the large rRNA precursor. In addition to the mature 18S, 5.8S and 25S (in yeast) rRNAs, it bears two external and two internal transcribed spacers (5' ETS and 3' ETS) and (ITS1 and ITS2), which are cleaved and processed in a complex process (Venema and Tollervey 1999; Fig. 6.3).

In *Saccharomyces cerevisiae*, the 35S rRNA precursor is first cleaved in the 3' ETS by RNAse III. The next cleavage is at site A0 in the 5' ETS (Hughes and Ares 1991) yielding a 33S precursor. This is subsequently processed at sites A1 and A2, giving rise to the 20S and $27SA_2$ precursors which are further processed to yield the larger mature rRNAs for the small and large ribosomal subunits, respectively. The 20S precursor is then processed to the mature 18S rRNA in the cytoplasm by clevage at its 3' end (Udem and Warner 1972; Veldmann et al. 1980). The $27SA_2$ is processed to mature 5.8S and 25S rRNA by one of the alternative processing pathways in ITS1 (Henry et al. 1994) and subsequent removal of ITS2. (see also schematic processing pathway in Fig. 6.3).

Several *trans*-acting factors have been identified implicated in pre-rRNA processing. Three snoRNAs, i.e U3, U14, SnR30 and their snRNP components Nop1p, Sof1p and Gar1p have been shown to be essential for 18S rRNA generation. Genetic depletion of any of these components leads to inhibition of the early cleavage events and to the insufficient accumulation of mature 18S rRNA. (Li et al. 1990; Hughes and Ares 1991; Tollervey et al. 1991; Girard et al. 1992; Jansen et al., 1993; Morrisey and Tollervey 1993).

Several of the RNA/RNA interactions take place between the essential snoRNAs and the pre-rRNAs. However, the relationship between this binding and the cleavage reactions is still unclear. Recent work on the U3 snoRNA is shedding new light on this phenomenon. U3 is the most abundant snoRNA present in the nucleolus (2×10^5 copies per cell). Depletion of U3 or its associated proteins, Nop1p and Sof1p leads to the inhibition of processing at sites A0, A1 and A2 as well as 18S rRNA synthesis (Beltrame et al. 1994; Beltrame and Tollervey 1995). It has previously been shown that a single-stranded region within U3 (termed box

A) can establish a perfect Watson-Crick base pairing interaction with the 5' ETS of the 35S pre-rRNA (Beltrame and Tollervey 1992; Beltrame et al. 1994). This base pairing is required for cleavages at sites A0 (100 nts), A1 (200 nts) and A2 (2000 nts) away from the transcription start site (Beltrame et al. 1994). *In vivo* crosslinking studies on U3 indicate that the phylogenetical conserved Box A region may be complementary to the universal conserved pseudoknot structure present at the very 5' end of the mature 18S rRNA (Hughes 1996).

Analysis of substitution mutations in 18S rRNA which disrupt the U3/18S interaction should abolish cleavage at sites A1 and A2 and prevent synthesis of mature 18S rRNA. We are currently investigating whether compensatory mutational changes in U3 which restore base pairing largely restore 18S synthesis, to show that this RNA/RNA interaction is required for the cleavage site recognition. This interaction is not required for the RNAse III-mediated cleavage at site A0, in contrast to the U3/5' ETS interaction. We therefore suggest that U3 interact at two distinct sites with pre-rRNA. It may further act as an RNA chaperone to establish the correct pre-rRNA structure required for recognition and processing events to take place.

It is believed that processing events in the 5' ETS and ITS1 are also coupled. Depletion of the essential snoRNAs U14 and snR30 lead to the accumulation of an aberrant precursor species, termed 23S, which extends from cleavage site A0 to A3 in ITS1, whereas depletion of U3 additionally also leads to the inhibition of cleavage at site A0. Mutations that delay or inhibit cleavage at site A1 lead to the accumulation of a 22S-precursor species, i.e. mutations at A1 concomitantly inhibit cleavage at A2 and processing continues at the downstream cleavage site A3. This suggests that cleavage at A1 precedes A2 (Venema et al. 1995). However, processing intermediates, which have been processed at the cleavage site A2 but not at A1, are not detected and indicate that A1 and A2 are coupled. It is thus postulated that a large processing complex containing snoRNPs assembles on the 5' ETS and brings all the cleavage sites in close proximity to each other so that processing events at sites A0/A1 and A2 can take place in a concerted manner (Morrisey and Tollervey 1995)

Interestingly, analysis of mutations at site A1, the 5' end of the 18S rRNA have established that its position is identified by two independent recognition elements (Venema and Tollervey 1999). The first involves the recognition of six phylogenetically conserved nucleotides around A1 (sequence-specific recognition). The second is governed from within the evolutionary conserved stem-loop structure, at which cleavage always occurs at a fixed distance from the base of the stem (spacing mechanism). Mutations, which affect the stability of the stem loop, should weaken the base-pairing interaction specifically, leaving the sequence recognition unaffected, in accordance with our data. Conversely, mutations, which alter the phylogenetical conserved nucleotide, which flanks A1, knock out the sequence-specific recognition without affecting the other mechanism. Neither, however, affects the overall cleavage efficiency of the processing site A1, which is in marked contrast to the U3/18S loop interaction in which both mechanisms are concomitantly inhibited. This seems to indicate that the stem and loop sequences of the 18S rRNA are also independently recognized during A1 cleavage, adding to the complexity of the early processing events taking place during ribosome biogenesis.

Small Nucleolar RNAs

Many snoRNAs have been identified in eukaryotic cells (Tollervey and Kiss 1997; Fig. 2.15). However, most of them are not required for cell viability, and thus their role has until recently, remained unclear. The non-essential snR31 influences processing of the primary transcript (Dandekar and Tollervey 1993; Fig. 2.15), but this may actually be explained by guide snoRNA function of this RNA (Kiss-László et al. 1996; Ni et al. 1997). A major breakthrough was the discovery that the snoRNAs can be generally assigned into two different classes based on their conserved sequence elements (Balakin et al. 1996). The families are classified as the box C/D and ACA class of snoRNAs respectively. The snoRNAs of the box C/D class are all associated with the nucleolar protein fibrillarin (Kiss-Lazlo et al. 1996; Tycowski et al. 1996). It has been shown that they serve as guide RNAs in site-specific ribose methylation of pre-rRNA (Cavaillé et al. 1996; Kiss-Laszlo et al. 1996; Tycowski et al. 1996, Bachellerie and Cavaillé 1997). The hallmark of these snoRNAs is the presence of an extensive sequence complementarity (10-21 nts) to the pre-rRNA. Base pairing between the snoRNA and pre-rRNA positions the conserved box D (terminal CUGA) or D' (internal CUGA) of the snoRNA precisely 5 bp away from the nucleotide to be methylated in the pre-rRNA (Fig. 2.16). Using a U24 snoRNA construct, Kiss-Lazslo et al.(1996) have shown that insertion of one nucleotide just upstream of the box D displaces the site of methylation by exactly one nucleotide, further confirming the model. The methyltransferase probably interacts with the D or D' box of the snoRNA and requires the recognition of the snoRNA/pre-rRNA complex for site-specific selection of the nucleotide to be modified (Cavaillé et al. 1996; Kiss-Lazlo et al. 1996). This appears to be conserved throughout evolution, as human U25 can rescue a methylation defect in *Xenopus* oocytes, as shown by Tycowski et al. (1996). Interestingly many of the snoRNAs are intron-encoded (Seraphin 1993).

The ACA class of snoRNAs is the second major class (Balaikin et al. 1996). All the members of this class can be folded into a universal hairpin-hinge-hairpin-tail secondary structure (Ganot et al. 1997; Fig. 2.15). They contain a set of conserved sequence elements, notably an ACA triplet located 3 nts before a conserved box H (consensus, AnAnnA) in the hinge region (Balakin et al. 1996; Ganot et al. 1997). The function of this class of snoRNas was poorly understood until the recent discovery of the "methylation guide" snoRNA. This prompted the speculation that the ACA class of snoRNA might also function in modification reactions of the pre-rRNA, notably pseudouridylation. Pseudouridylation is another major modification that the pre-rRNA undergoes after transcription, in which uridines (95 in vertebrates and 45 in yeast) are converted to pseudouridines (Maden 1990).

Interestingly, these modifications are highly conserved during evolution and show patterns of clustering in the pre-rRNA. Work done in the laboratories of T. Kiss and M. Fournier has established that the ACA class of snoRNAs is indeed required for site-specific pseudouridylation of the pre-rRNA. Depletion of specific snoRNAs in yeast had been shown to abolish site-specific pseudouridylation of pre-rRNA, which can be rescued upon reintroduction of the corresponding wild-type genes (Ganot et al. 1997; Ni et al. 1997). The nucleotide, which is to be modified, is determined by snoRNA/pre-rRNA interactions, akin to the box C/D class of snoRNAs. Inspite of low sequence complementarity (as compared to box C/D snoRNA/pre-rRNA) Ni et al. (1997) showed that distinct structural elements are conserved among the individual snoRNAs. Two domains, termed A and B, show a short (5-9 nts) region of uninterrupted complementarity to the pre-rRNA, in which the pseudouridine site is always located immediately adjacent to the domain A or 1 nucleotide away. In contrast, the ACA box is always positioned at a fixed distance of 15 nts from the site of pseudouridylation (Fig. 2.15). Disruption of this base-pairing leads to the abrogation of pseudouridine formation. Ganot et al. (1997) have also shown that the snoRNAs can fold into a common secondary structure (hairpin-hinge-hairpin-tail). The site of pseudourinylation is flanked by two short rRNA sequences, which base pair to the snoRNA. The modification site occupies the first unpaired position following the helical structure. The distance between the box ACA and the internal loop with the "pseudouridine-pocket" is always 15 to 16 nts (Fig.2.15).

Gar1p, which is a nucleolar protein common to all the ACA class of snoRNAs has also been shown to be involved in the modification process as depletion of the protein leads to the global loss of pseudouridylation (Bousquet-Antonelli et al. 1997). However, Gar1p, does not show any homology to any known pseudouridine synthase and thus is most likely not to be the enzyme.

Thus, it has been established that a bipartite recognition signal, which involves two short stretches of complementarity between the snoRNA and pre-rRNA, operates to determine the uridine, which is to be modified. The role of the modifications is as yet unclear, as the lack of individual snoRNAs responsible for these reactions does not have any effect on yeast growth (Parker et al. 1988). It has been postulated that modifications might affect the overall stability of the rRNA and aid proper folding of the rRNA during ribosome biogenesis.

Modified nucleotides occur also as micromotifs (only one modified nucleotide, yet with a particular function) in snoRNAs, probably for structural or stability reasons and it is suggested that also for these guide snoRNAs exist.

Ribonuclease E (RNase E) from *E. coli* plays a central role in the processing of rRNA, degradation of mRNA and the control of replication of ColE1-type plasmids. Surprisingly, Huang et al. (1998) found that RNAse E is able to shorten 3' poly(A) and poly(U) homopolymer tails on RNA molecules. This suggests a mechanism by which Rnase E may exercise overall control of mRNA decay. Mackie (1998) showed that Rnase E frequently proceeds in a 5' to 3' direction and that RNase E cleaves natural substrates poorly with 5' -triphosphate groups, whereas 5' mono-phosphorylated substrates are strongly preferred. This indicates that RNase E has inherent vectorial properties with its activity depending on the 5' end of its substrates. Diwa et al. (2000) mention that an evolutionary conserved RNA stem-loop functions as a sensor that directs feedback regulation of RNase E gene expression. Autoregulation of RNase E is mediated in *cis* by the 361 nts 5' UTR of RNase E mRNA.

6.7   Nuclear RNA Transport

Import and export of RNA into and out of the nucleus involves many different RNA motifs (Görlich and Mattaj 1996). The transport of the RNA during its maturation from precursor in the nucleus to mRNA in the cytoplasm is regulated by recognition of specific motifs, a simple and well-characterized example is the poly(A) tail involved in labelling mRNA as mature and ready for transport into the cytoplasm (Wickens et al. 1997). Further different examples of pre-mRNA transport, including export of viral mRNAs, are also a focus of research. Important research questions address the specificity of the RNA signals as well as the specificity of the proteins recognizing them, the interaction of the RNA with the nuclear pore complex, and the transport of specific classes of RNA by different nucleoporins. For example, Nup145 binds RNA homopolymers (e.g. poly G). During heat shock there is an emergency system for specific transport of heat shock related mRNAs. Different nucleoporin mutants defective in pre-tRNA transport have been isolated. Mutations in conserved regions of pre-tRNA lead to transport defects, pointing to a specific RNA motif recognized here. Nucleotide modifications constitute important micromotifs specifically recognized by interacting proteins, for instance the m_3G cap on small nuclear RNAs. Several nuclear RNA viruses face a common problem: they are spliced to yield several different proteins during their replication cycle, yet the unspliced, complete RNA must be exported from the nucleus to pass the complete genomic information to the next generation. HIV solves this problem by the rev-response element recognized by the rev protein synthesized by the virus itself. Rev achieves export of the complete HIV RNA. However, murine moloney leukemia virus achieves this by a different regulatory RNA element, the cytoplasmic transport element (CTE) first recognized in Mason Pficer monkey virus (MPMV). The CTE is recognized by a host protein, the human RNA helicase I. Saturation with CTE blocks mRNA export and to a much lesser extent snRNA export from the nucleus, but not tRNA export. In contrast, saturating the distinct transport system targeted by the rev protein blocks the export of snRNA, but not of mRNA. The different 3' end of histone mRNA is indicative of yet another transport system for this type of messenger.

6.8   Regulatory Motifs in mRNA

The 5' Untranslated Region in mRNA

Iron-responsive elements (IRE) represent an example of current research on regulatory RNA motifs in the 5' UTR. These are regulatory RNA elements, which are characterized by a phylogenetic defined sequence-structure motif (Fig. 4.1). Their biological function is to provide a specific binding site for the IRE-binding protein (IRP). Iron starvation of cells induces high-affinity binding of the cytoplasmic IRP to an IRE, which leads to repression of ferritin mRNA translation. Using a combination of database searching, different screening filters and experimental tests (Chaps. 3, 4), we found several new functional IREs (Dandekar et al. 1991; Dandekar and Hentze 1995). Examples concern translational regulation of mammalian and *Drosophila* citric acid cycle enzymes (Gray et al. 1996). *D. melanogaster* IRP binds to an IRE in the 5' UTR of the mRNA encoding the iron-sulfur protein (Ip) subunit of succinate dehydrogenase (SDH). Interestingly, this interaction is regulated resulting in different translation of this mRNA in *Drosophila* embryogenesis. In a cell-free translation system, recombinant IRP-1 imposes specific translational repression on a reporter mRNA bearing the SDH IRE, and the translation of SDH-Ip mRNA is iron-regulated in *D.melanogaster* Schneider cells.

In mammals, an IRE was identified in the 5' UTR of mitochondrial aconitase mRNA from two species. Recombinant IRP-1 represses aconitase synthesis with similar efficiency as ferritin IRE-controlled translation. The interaction between mammalian IRPs and the aconitase IRE is regulated by iron, nitric oxide, and oxidative stress (H_2O_2), indicating that these three signals can control the expression of mitochondrial aconitase mRNA. These results identify a regulatory link between energy- and iron metabolism in vertebrates and invertebrates and suggest biological functions for the IRE/IRP regulatory system in addition to the maintenance of iron homeostasis.

To complement previous search results for new IREs in eukaryotes, a systematic analysis of prokaryotic sequences is in progress. The consensus IRE motif demands a specific stem-loop structure with a second helix attached after a bulged C and several obligatory nucleotides in the loop region. Negative filters exclude incompatible nucleotides and unstable secondary structures. Candidate mRNA structures satisfying these criteria are next examined for correct position within the mRNA. The remaining RNA structures are checked in detail for biological evidence implicating involvement in iron metabolism. Only very few prokaryotic RNA structures remain after all filters; three different structures are currently being investigated. However, it should be kept in mind that the consensus of the IREs functional in eukaryotes is comparatively strict, and that in prokaryotes wider variations of the motif may be used.

In activated T cells Interleukin-2 (IL-2) mRNA is stabilized by signaling pathways. IL-2 mRNA contains at least two *cis* elements that mediate its stabilization in response to different signals, including activation of JNK. This is mediated through a *cis* element that encompasses the 5' UTR and the beginning of the coding region. Lacking this 5' element, IL-2 mRNA is still responsive to other signals that probably aim towards the 3' UTR (Chen et al. 1998).

Poly(A)-binding protein (PABP) mRNA can be autoregulated through a 61 nts long A-rich sequence in its 5' UTR. Overexpression of PABP down-regulates the translation or the abundance of its own mRNA (Hornstein et al. 1999).

Open Reading Frames (ORFs) in mRNA

As indicated in the conclusion of Chapter 4, different regulatory motifs, as well as co-translational protein folding requirements (Brunak and Engelbrecht 1996; Thanaraj and Argos 1996) overlap in the ORF of the mRNA. Doubtless more regulatory and functional circuits will be discovered. For illustration, the following section discusses two current examples.

Completeness of the cDNA Sequence

This means, in other words, whether the mRNA you have just cloned is complete or did you overlook important parts (segments, motifs, regulatory structures or coding region) at the ends? As expressed sequence tags and genome sequencing flourish, this question has immediate practical applications. The accumulated knowledge on ORF regulation and regulatory RNA motifs involved allows estimates for assessing the completeness of a cDNA sequence, including identification of the start site for translation (Kozak 1996). Basic features indicating the translation start are the ATG codon and a favorable context for initiation, the presence of an upstream in-frame terminator codon and the prediction of a signal peptide-like sequence at the amino terminus. However, examples from the literature illustrate the limitations of these criteria. It is best (see Chap. 4) to inspect a cDNA sequence not only for these positive features but also for the absence of certain negative indicators. Three specific warning signs for mistakes are:

- The presence of numerous AUG codons upstream from the presumptive start site for translation. This often indicates an aberration (sometimes a retained intron) at the 5' end of the cDNA: Thus, this sequence is not all RNA in reality or a mixed or mutated version.

- A strong, upstream, out-of-frame AUG codon poses a problem if this reading frame starting from the upstream AUG overlaps the tentative start of the major ORF. cDNAs that display such an arrangement often turn out to be incomplete: their out-of-frame AUG codon is within, rather than upstream of, the protein coding domain (re-sequence and extend it to get the complete coding sequence, let alone any 5' untranslated regulatory motifs which you definitely have missed in

such a shortened cDNA clone).
- A very weak context at the putative start site for translation often means that the real initiator codon has been missed, and again the mRNA with any regulatory signals and its ORF is only incompletely present.

Regulation of mRNA Translation by tRNA Redundancy

tRNA gene redundancy and translational selection in yeast has been examined in detail (Percudani et al. 1997). The authors aim was to better understand which set of tRNAs is chosen to interact with the mRNA coding for a high or low expressed protein and which structural requirements of the tRNA are needed for this. A dedicated algorithm, Pol3scan, (compare with Chap. 4) extracted all tRNA sequences. The entire tRNA set of yeast consists of 274 tRNA genes. The gene copy number for individual tRNA species ranges from 1 to 16 and correlates well (r=0.82) with the frequency of codon occurrence in a sample of 1756 distinct protein coding sequences and the previously measured intracellular content of 21 tRNA species. In particular, regression analysis values for individual protein-coding sequences proved to be an effective description of the translational selective pressure operating on a particular gene. Analyzing the structure of the different tRNAs identified, four deviations from previously proposed rules for third-position wobble pairing in yeast, three G:U and one A:I codon-anticodon pairings, were found in the whole set of yeast tRNAs translating the 61 coding triplets.
Analyzing the recognition of tRNA itself by charging tRNA synthetases, previous studies showed *E.coli* tRNAs to be dominantly recognized by tertiary structural elements. A study by Soma et al. (1996) on tRNA^{Leu} in yeast shows that mutating the second position of the anticodon and the 3' adjacent to the anticodon (positions 35 and 37) as well as position 73 (the discriminator base) was sufficient to convert tRNA^{Ser} into an efficient leucine acceptor. Even a one-nucleotide insertion in the variable arm of tRNA^{Leu} is sufficient to confer an efficient serine accepting activity (Himeno et al. 1997).

3' UTR in Parasites

Regulatory signals in the 3' UTR are also an upcoming research topic in parasitology. Thus the 3' end microheterogeneity in the poly(A)-tail of circum-sporozoite antigen, for instance in *Plasmodium berghei* seems to result from different signals around heterogeneous polyadenylation sites (Ruvolo et al. 1993). Similarly, varying levels of expression from reporter genes could be regulated by different portions of the 3' UTR from the hexose transporter mRNA in *T. brucei* (Hotz et al. 1995). In *Plasmodium gallinaceum,* the DNA sequence of the pgs28 3' gene-flanking region contains seven eukaryotic polyadenylation signals (AATAAA/ATTAAA) and a T-rich region 55 nts upstream of the fifth polyadenylation signal. Golightly et al. (2000) showed that the U-rich element and the fifth eukaryotic polyadenylation consensus sequence within the pgs28 3' UTR are necessary for Pgs28 protein expression.

3' UTR in Viridae

Alfalfa mosaic virus (AMV) coat protein and tobacco streak virus (TSV) coat protein bind specifically to the 3' UTR of the viral RNAs and are required with the genomic RNAs to initiate virus replication. Analyzing the determinants important for specific binding of the 3' terminal 39 nts of AMV RNA (AMV843-881) to an amino-terminal coat protein (CP26) revealed a series of potential contacts (Ansel-McKinney and Gehrke 1998). Nucleotides found important for RNA-protein interactions are highly conserved among AMV and TSV RNAs. This seems to be the explanation for the fact that heterologous AMV and iliavirus coat protein-RNA mixtures are infectious.

Seleno-Cysteine mRNA and their 3' UTR Motifs

In mammalian selenoprotein mRNAs the recognition of UGA as a seleno-cysteine (Se-Cys) codon requires Se-Cys insertion elements (SECIS) that are contained in a stable stem-loop structure in the 3' UTR (Fig. 2.4; Tormay et al. 1996; Berry 2000). Instead of a translational stop at the UGA codon, this directs incorporation of an appropriate Se-Cys. Such a modified cysteine residue contains the Se instead of sulfur. This is critical for the activity of several enzymes, for instance in redox protection such as the glutathione peroxidase (GPx) in many organisms including vertebrates and their unicellular parasites.

For prokaryotes a model of Se-Cys incorporation could be developed (Berry 2000). The Se-Cys tRNA is only poorly recognized by the standard elongation factor EF-Tu in prokaryotes (Baron and Boeck 1991) and eEF1A in eukaryotes (Jung et al. 1994). Instead, Se-Cys tRNA is bound by the Se-Cys specific elongation factor (SELB) (Forchhammer et al. 1989). SELB in prokaryotes possesses an amino-terminal domain with high homology to EF-Tu and a carboxy-terminal domain that recognizes the SECIS element, recruiting the factor to the UGA Se-Cys codon (Baron et al. 1993; Huttenhofer and Boeck 1996). In eukaryotes the identification of the SELB homologues has proved to be much more difficult.

Lesoon et al. (1997) investigated the SECIS elements and cellular proteins required for Se-Cys insertion in rat phospholipid hydroperxide glutathione peroxidase (PhGPx). Interestingly, the catalytic activity of this enzyme and its protection ability against oxidative stress critically depends on Se-Cys incorporation. From rat testis a 120 kDa protein binds specifically to the PhGPx and GPx mRNA 3' UTR. Binding of the 120 kDa protein requires the AUGA SECIS element; but it is not clear if this protein plays a role in selenoprotein synthesis (Berry 2000). Recently, a mammalian SECIS-binding protein, designated SBP2, was purified and cloned by Copeland et al. (2000). SBP2 exhibits specificity for wild-type, but not mutant SECIS elements and functions to enhance Se-Cys incorporation *in vitro*, although it does not contain an EF-homologous domain. Rother et al. (2000) report that Se-Cys insertion into archeal seleno-polypeptides is directed through the SECIS element in the 3' UTR as in eukaryotes.

Detailed Analysis of a Complex RNA Stability Motif in the 3' UTR

The number of regulatory RNA structures in untranslated RNA is probably far higher than previously anticipated; however, many of these can only be revealed by a close analysis of different RNA foldings. Recent examples include decay of mRNAs containing a non-sense mutation involving *cis*-acting mRNA elements with a loose consensus sequence positioned downstream of the termination codon and at least three *trans*-acting proteins in yeast. A similar system is found in *C.elegans* and there is some indication that this mRNA controlling mRNA protein interactions exists also in humans. Or, for instance, the *cis*-acting RNA signals within the 3' UTR known as the cytoplasmic polyadenylation element (CPE), involved in mRNA polyadenylation and translational activation during oocyte maturation (Wickens et al. 1997, 2000). As an example from our own work we mention the analysis of the 3' UTR of human oestrogen receptor mRNA (Kenealy et al. 1996). It contains a signal mediating rapid degradation of the mRNA, but this is only apparent if different sub-segments of the mRNA and higher-order foldings are compared. This highlights the fact that regulatory RNA structures may also depend on long-distance RNA interactions. Furthermore, such an approach is an alternative to reveal unique and complex regulatory RNA motifs without reference to a known template structure. Thus, this region has the ability to reduce mRNA levels and probably acts as an instability signal (Sachs et al. 1997). To tackle the RNA structure, again theoretical and experimental approaches have to be combined. Alternative RNA foldings are theoretically analyzed to identify regions important for RNA stability. Transfection assays with different constructs subsequently test the effect of different structures on the mRNA stability.

Polyadenylation signals

The mammalian sequences involved in polyadenylation have a broad consensus of AAUAAA, followed by a CA, cleavage site, followed by a GU sequence (more details see above). At first sight the yeast sequences around the polyadenylation signal look quite different. There is no conservation of an AAUAAA sequence. It starts with an AU-rich region, followed by an A-rich region; cleavage occurs after pyrimidine A. However, despite conflicting views on single proteins, there is now consensus in the field (Keller 1995) that the proteins from mammals involved in polyadenylation CPSF (polyadenylation site factor with four proteins 160, 100, 73 and 30 kDa), CstF (cleavage-stimulating factor, acting downstream of the cleavage site with proteins of 77, 64 and 5 kDa) and, in between CFI, CFII and PAP (poly(A) polymerase) have mostly homologues in yeast and that the difference in the recognition requirements is only superficial. The overall signal-processing machinery is the same despite divergence in some sequence details and requirements in the poly(A) region of the mRNA. Klasens et al. (1999) showed that the ability of the HIV-1 AAUAAA signal to bind polyadenylation factors is controlled by local RNA structure. This structure is termed poly(A) hairpin because it contains polyadenylation signals, the cleavage site and part of the GU-rich downstream element.

Developmental Differentiation Signals in the 3' UTR

Increasingly, developmental signals are found to be encoded by RNA motifs, for instance in the 3' UTR of an mRNA. *C. elegans* switches from being male to subsequently female. Isolation of gain of function mutants reveals that the gene *fem 3* (female 3) leads to worms which produce sperm only by mutations in the 3' UTR of this mRNA (Ahringer and Kimble 1991). The RNA structure responsible for this has been mapped in 17 gain of function mutants to a small point mutation element, *pme* (UCUUG). To isolate protein factors involved in RNA translational control during development, Wickens and his group developed in collaboration with Stanley Fields a three-hybrid assay system. It consists of a lexA DNA-binding domain on one construct fused to an RNA-binding protein, in this case the MS2 coat protein, the RNA under investigation, in this case the pme RNA fused to the MS2-binding site on the next construct. The third construct provides the Gal4 activator domain fused to a cDNA library (SenGupta et al. 1996). Thus, Gal4 fusions test, which isolated protein can provide bridging to the *pme*-RNA by efficient binding to it.

FBF1 protein could be identified by this three-hybrid assay for RNA-protein interactions. Ablation of FBF1 mRNA using antisense technology yields animals that produce only sperm. *Fem-3* loss of function leads to worms that have only oocytes. As the cross between both mutants leads to obligatory female, fem3 acts downstream of FBF1. The FBF1 protein has homology to the *Drosophila* protein *pumilio,* which inhibits *hunchback* mRNA in *Drosophila*. Comparing both processes suggests that the regulatory pathway operates by deadenylation of the downstream mRNA, respectively *hunchback* in *Drosophila* and fem 3 in *Caenorhabitis*, the complex 3' structure of both RNAs being the critical regulator. This is not known yet; however, in FBF1 loss of function mutants, *fem-3* mRNA has a long poly(A) tail.

Katsu et al. (1999) studied the regulation of cyclin B mRNA by 17-alpha,20-beta-dihydroxy-4-pregnen-3-one (ABDP, maturation-inducing hormone) during oocyte maturation in the goldfish. They showed that cyclin B mRNA is present in both immature and mature oocytes. Sequence analysis revealed that goldfish cyclin B mRNA contains four copies of cytoplasmic polyadenylation element (CPE)-like motifs in the 3' UTR, suggesting that the initiation of cyclin B synthesis during oocyte maturation may be controlled by elongation of poly(A)-tail. Cordycepin, an inhibitor of poly(A) addition of mRNA, prevented ABDP induced oocyte maturation. These findings suggest that in goldfish oocytes, the synthesis of cyclin B protein is under translational control and that cytoplasmic 3' poly(A) elongation is involved in ABDP-induced translation of cyclin B mRNA.

Ralle et al. (1999) characterized *cis*-acting sequences of *Xenopus* nuclear lamin B1 mRNA that mediate translational regulation. They show that two CPE-like elements in the 3' UTR of B1 mRNA act as translational repressors in oocytes.

In *Xenopus* oocytes, progesterone interacts with an unidentified surface-associated receptor, which induces a non-transcriptional signaling pathway that results in a cytoplasmic readenylation and translational activation of a selected group of maternal mRNAs, including *mos, cyclinA1* and *B1*, and *cdk2*. Their translational recruitment is regulated by cytoplasmic polyadenylation. This process requires two *cis*-acting 3' UTR elements and three core polyadenylation factors. The *cis* elements are the hexanucleotide AAUAAA, and an upstream U-rich sequence called cytoplasmic polyadenylation element (CPE) (Richter 1999). The core factors include CPEB, an RNA recognition motif and zinc finger-containing protein that binds the CPE (Hake and Richter 1994; Hake et al. 1999), and a cytoplasmic form of cleavage and polyadenylation specificity factor (CPSF), which interacts with AAUAAA (Dickson et al. 1999). The third core factor is poly(A) polymerase (PAP) (Gebauer and Richter 1995). Mendez et al. (2000a) define the molecular function of CPEB in cytoplasmic polyadenylation. The ordered events that comprise this process begin with the binding of progesterone to an as-yet- unidentified cell surface receptor, which is followed by a transient decrease in cAMP and the activation of the kinase Eg2 (Andresson and Ruderman 1998). Active Eg2 then phosphorylates mRNA-bound CPEB on Ser 174, an event that enhances the interaction of CPEB with the CPSF 160 kDa subunit. This interaction most likely stabilizes the binding of CPSF to the AAUAAA, which is important for the recruitment of PAP to the mRNA.

Developmental Localization Signals in the 3'UTR

Regulatory motifs in *oskar* mRNA are central to recruit other maternally provided components required to form a functional pole plasm (Ephrussi and Lehmann 1992; Wickens et al. 2000). Pattern formation in early development relies on localized cytoplasmic proteins, which can be prelocalized as mRNAs. This effect was revealed by *in situ* hybridization for posterior pole localization of *oskar* mRNA in the *Drosophila* oocyte (Kim-Ha et al. 1991). Specific RNA regulatory elements in the 3' UTR direct mRNA localization as demonstrated using a hybrid lacZ/*oskar* mRNA, and serial deletions (Kim Ha et al. 1993). Pokrywka and Stephenson (1995) used different inhibitory drugs to show that microtubules are important in anterior localization of *bicaudal-D*, *fs(1)K10* and *orb RNA* whereas *bicoid* RNA localization is less and *oskar* RNA is not dependent on this but on cytoplasmic tropomyosin (Erdelyi et al. 1995). Serano and Cohen (1995) looked at the transport of *K10* mRNA from nurse cells into the oocyte. The authors used a transgenic fly assay to analyze the expression patterns of a series of *K10* deletion variants. They identified a 44 nts transport/localization stem-loop sequence (TLS), a signal also present in *orb* mRNA. Translational control of *oskar* generates short OSK, the isoform that induces polar granule assembly (Markussen et al. 1995). Prior to its localization translation of the *oskar* mRNA has to be repressed. Translational repression is mediated by an ovarian protein, *bruno*, that binds specifically to *bruno* response elements (BREs), present in multiple copies in the *oskar* mRNA 3' UTR (Kim-Ha et al. 1995). Addition of BREs to a heterologous mRNA renders them sensitive to translational repression in the ovary.

Staufen is also transported with *oskar* mRNA during oogenesis. It associates specifically with both *oskar* and *bicoid* mRNAs to mediate their localization, but at two distinct stages of development. *Staufen* protein is required to anchor *bicoid* mRNA at the anterior pole of the *Drosophila* egg. The protein co-localizes with *bicoid* mRNA at the anterior pole, this localization depends on the association with the mRNA (Ferrandon et al. 1994). Using embryo injection and deletion analysis the sequences required could be mapped to three regions of the 3' UTR, each of which is predicted to form a long stem-loop. The resulting *staufen-bicoid* 3' UTR complexes form particles that show a microtubule dependent localization.

Another example for such a network of interacting developmental genes and 3' UTR regulation is *c-myc* mRNA. It bears a localization signal in the 3' UTR trageting it to the perinuclear cytoplasm and cytoskeletal-bound polysomes. The 3' UTR can be linked to other mRNAs and directs these to the same locations. Bases 194-280 of the UTR seem to carry this function, a conserved AUUUA element seems to be critical (Veyrune et al. 1996).

6.9 Posttranscriptional Gene Silencing and RNA Interference

Posttranscriptional gene silencing (PTGS) and RNA interference (RNAi) refer to similar mechanisms. PTGS is found in plants. Here, it describes a defence mechanism of degrading mRNAs. Dalmay et al. (2000) describe four gene loci required for PTGS in *Arabdidoposis*. One of these, SDE1, encodes an RNA-dependent RNA polymerase; its role is thought to synthesize a double-stranded RNA initiator of PTGS. This RNA-dependent RNA polymerase is required by transgenes but not by viruses; here the virus-encoded RNA polymerase produces double-stranded RNA. Mourrain et al. (2000) isolated sgs2 and sgs3 *Arabidoposis* mutants impaired in PTGS and showed that PTGS is an antiviral defence mechanism that can also target transgene RNA for degradation. Hamilton and Baulcombe (1999) furthermore showed that antisense RNA complementary to the targeted mRNA was detected. These RNA molecules were of a uniform length, estimated at 25 nts.

RNAi is found in animals, and here double-stranded RNA (dsRNA) directs the sequence-specific degradation of mRNA (Sharp 1999). Zamore PD et al. (2000) developed a *Drosophila in vitro* system in which they showed that RNAi is ATP-dependent yet uncoupled from mRNA translation. During RNAi reaction both strands of the dsRNA are processed to RNA segments 21-23 nts in length. The mRNA is cleaved only within the region of identity with the dsRNA. Tabara et al. (1999) selected *C. elegans* mutants resistant to RNAi. Two loci, rde-1 and rde-4, were defined by mutants strongly resistant to RNAi but with no obvious defects in growth or development. Several, but not all, RNAi-deficient strains exhibit mobilization of the endogenous transposons. The authors discuss the possibility that one natural function of RNAi is transposon-silencing. The genetic requirements for inheritance of RNAi in *C. elegans* were investigated by Grishok et al. (2000). They found that transmission of RNAi occurred through a dominant extragenic agent. Rde-1 and rde-4 were required for the formation of this interfering agent but were not needed for RNAi thereafter. Rde-2 and mut-7 genes were required for downstream interference. Ketting and Plasterk (2000) found that mutants of *C. elegans* that are defective in transposon silencing and RNAi (mut-2, mut-7, mut-8 and mut-9) are, in addition, resistant to co-suppression. This indicates that RNAi and co-suppression in *C. elegans* may be mediated at least in part by the same molecular machinery, possibly through degradation of mRNA molecules. Hammond et al. (2000) generated 'loss-of function' in *Drosophila* cell Hurenkinders that were transfected with specific dsRNA. That coincides with a marked reduction in the level of cognate mRNA. Extracts of transfected cells contained a nuclease activity that specifically degraded exogenous transcripts homologous to transfected dsRNA. This enzyme contained an essential RNA component and it may confer specificity to the enzyme through homology to the substrate mRNA. Domeier et al. (2000) tested how RNAi is affected by mutations in the *smg* genes, which are required for nonsense-mediated decay (NMD). For three of six *smg* genes tested, mutations resulted in animals that were initially silenced by dsRNA but they recovered, suggesting that persistence of RNAi relies on a subset of *smg* genes. Braun (2000) reports that during mammalian spermatogenesis temporal translational regulation of protamine 1 (Prm 1) mRNA is dependent on a highly conserved sequence located in the distal region of its 3' UTR. The 17-nt translational control element (TCE) mediates translational repression of the Prm1 mRNA. Mutation of the TCE causes premature synthesis of protamine protein and sterility. Proper translational activation of the Prm1 mRNA in elongated spermatides requires the cytoplasmic double-stranded RNA-binding protein TARBP2. TARBP2 is expressed at low levels in many cells, but at high levels in the late-stage meiotic cells and in postmeiotic spermatides. Mice, mutant for TARBP2, are defective in proper translational activation of the Prm1 and Prm2 mRNAs, and are sterile. The NMD pathway functions to degrade transcripts containing non-sense codons. Transcripts containing mutations that insert an uORF in the 5' UTR are normally degraded; but some uORF-containing transcripts are resistant to degradation. Ruiz-Echevarria and Peltz (2000) found that GCN4 and YAP1 mRNAs contain uORFs that harbour a stabilizing element (STE) that prevents rapid NMD by interacting with the RNA-binding protein Pub1. Therefore Pub1p is a critical factor that modulates the stability of uORF-containing transcripts.

Andersson et al. (2000) report two mammalian cytosolic proteins (BARB1 and BARB2) that selectively interact with the 3' UTR of the mRNA coding for the catalytic beta-subunit of mitochondrial ATP synthase (beta-mtATPase) in a novel and interdependent manner. They suggest that the interaction of BARB1 and BARB2 and beta-mtATPase mRNA involves the formation of a complex that may be involved in post-transcriptional regulation of gene expression.

6.10 Catalytic RNA and Motifs

RNAse P and catalytic group II introns are just two examples for recent research in the field of catalytic RNA; others have been mentioned already such as those generated by different SELEX approaches (Chap. 3) and group I introns.

RNAse P

The catalytic activity of RNAse P rests principally in the RNA component (Fig. 2.23). A tertiary fold, namely an RNA cage forms around an internal tRNA-like fold and attacks different RNA substrates, for instance tRNA and 4.5S RNA (Fig. 2.24). pC4 RNA and M3 RNA are further alternative substrates for RNAse P. Current research tries to better understand the interaction and reaction mechanisms between RNAse P and its substrates. In a genetic study, Kufel and Kirsebom (1996) studied the RNA for *E.coli* RNAse P, which is called M1 RNA in *E.coli*. Residues important for cleavage site selection and divalent metal ion binding were examined. The P7 loop from RNAse P and the 3' terminal RCCA motif in the tRNA substrate are critical for cleavage site selection as confirmed by cross-linking data on precursors. Base-pairing between the catalytic M1 RNA and the tRNA substrates results in a re-coordination of divalent metal ions such that cleavage at the correct position is accomplished. The P15 miniloop (34 nts) in the M1 RNA with its GGU is specific for positioning and metal binding. Adding protein C5 to the catalytic M1 RNA makes it more specific, faster and stable. This is true also in several other cases (but not always) where RNA enzymes are not as specific as the "newer" protein enzymes.

Two specific mechanisms direct RNAse P cleavage:

1) A fixed distance in the tRNA from the T loop to its 3' end is recognized.

2) RNAse P recognizes the 3' terminal CCA of the tRNA.

Both mechanisms are easily separable in an artificial tRNA substrate containing a stem extended by three nucleotides. In this case, +1 cleavage at the CCA terminus of the tRNA is observed but also at +4, in a fixed distance from the T loop of the tRNA. Phylogenetic and deletion studies show that about 260 nts are minimally required for the catalytic RNA in RNAse P to be active. Spatial proximity of parts far away in the sequence is required and prevents further deletion variants of the RNA to be active.

Warnecke et al. (1996) point to unique features of the transition state geometry in the RNAse P RNA-catalyzed reaction, converting RNAse P RNA into a cadmiumion-dependent (Cd^{2+}) ribozyme cleaving a modification (Rp-phosphorothioate) in the precursor tRNA at the RNAse P cleavage site.

Stams et al. (1998) determined the crystal structure of *Bacillus subtilis* RNAse P protein at 2.6 Å resolution. The unusual left-handed beta-alpha-beta connection to the RNA cleft topology is partly shared with ribosomal protein S5 and the ribosomal translocase elongation factor G, which suggests evolution from a common RNA-binding ancestor in the primordial translation apparatus.

Self-Splicing Introns

Central domains in self-splicing introns are domain I, providing EBS and IBS interaction, epsilon-epsilon' interactions and gamma-gamma' interaction; the domain V for catalysis and domain VI for the branch point. However, more subtle interactions in the other domains are now amenable to study, for instance in domain II, easily recognizable from its secondary structure. Though this domain is not essential, its conservation points to a biological function. Multiple tertiary interactions involved in domain II of group II self-splicing introns were examined by Costa et al. (1997). Though domain II (from a total of six) was long thought to be relatively unimportant for group II self-splicing, novel tertiary interactions have been found. t-t' is a novel tertiary interaction between the terminal loop of the IC1 stem of domain I and the basal stem of domain II and appears to stabilize the ribozyme core, and further interactions exist between domain II and VI. Regions essential for catalytic activity are located in domains V and I. They are linked by tertiary interactions (lambda-lambda') that create a framework for an active site and anchor it at the 5' splice site. Highly conserved elements similar to the lambda-lambda' interaction are also present in the eukaryotic spliceosom. (Boudvillain et al. 2000). Zimmerly et al. (1999) showed that yeast mitochondria group II introns encode reverse transcriptases that function in both intron mobility and RNA splicing and bind specifically to unspliced precursor RNA to promote splicing. Then they remain bound to the excised intron and form a DNA endonuclease that mediates intron mobility by target DNA-primed reverse transcription. Reverse transcriptase activity encoded by the yeast mtDNA group II intron aI2 is associated with an intron-encoded protein of 62 kDa (p62). This is bound tightly to endogenous RNAs in mtRNP particles, and the reverse transcriptase activity is rapidly and irreversibly lost when the protein is released from the endogenous RNAs by RNase digestion.

6.11 Viral RNAs

New viral RNAs are constantly being characterized, for instance variants of the HIV virus (Cullen and Malim 1991), where a large number of studies by different laboratories have been conducted on regulatory motifs such as the TAT-RNA, the rev-response element and the TAR HIV-1 leader RNA. Erard et al. (1998) found that the cellular protein TRBP binds and perhaps modifies TAR by the means of a "2-G hook" motif. The binding specificity of this double-stranded RNA-binding protein lies within the highly conserved dsRBD core motif. Analyzing involved regulatory RNA motif side effects of HIV infection may impair human mRNAs if they use similar signals. After applying the methods outlined in Chapter 4, we found an RRE-like structure in human DNA ligase I mRNA sequence which is phylogenetically conserved (Dandekar and Koch 1996) and may interact with HIV rev protein or a human rev-like protein. Crosslinking and competition data suggest that there is a specific cellular partner for the motif; however, the binding specificity seems to differ in some extent from HIV rev protein. Viridae are selected for swift replication and high copy number and hence for a compact genome (Table 2.3) with a high density of RNA motifs (Chap. 5).

With the tables in Chapter 2 and tools outlined in Chapters. 3 and 4, the reader should be able to identify important regulatory RNA motifs in a particular viral RNA he or she is studying. The following example recapitulates some of the steps involved: alternative structures of the cauliflower mosaic virus 35S RNA leader were identified by Hemmings-Mieszczak et al. (1997). This 600 nt leader contains regulatory elements involved in splicing, polyadenylation, translation, reverse transcription and probably also packaging. The authors predicted a conformation of a low-energy elongated hairpin, base-pairing the two halves of the leader, with a cross-like structure at the top, and confirmed this structure by enzymatic probing, chemical modification and phylogenetic comparison (Chap. 4). The long hairpin is stabilized by strong base-pairing between the ends of the leader, regions which are important in allowing translation downstream of the leader via the ribosome shunt mechanism. Interestingly, at high ionic strength the 35S RNA leader exhibits additional higher order structures of low electrophoretic mobility: a long-range pseudoknot connecting central and terminal parts of the leader and, secondly, a dimer. Thus, the authors could identify alternative functional RNA structures, probably involved in regulating cauliflower mosaic virus life cycle.

RNA synthesis that is protein-primed has been discovered for purified poliovirus RNA polymerase. A small protein, VPg is covalently linked to the 5' terminal uridylic acid of the plus-stranded poliovirus genomic RNA and is uridylated by the poliovirus RNA polymerase 3Dpol. This uridylated VPg can then prime the transcription of polyadenylated RNA by 3Dpol to produce VPg-linked poly (U). Initiation of transcription of the poliovirus genome from the 3' end therefore depends on VPg (Paul et al. 1998).

Perspective: the Increasing Set of RNA Motifs

Several, newly identified types of RNA underline the rapid growth of known RNA types and contained regulatory RNA motifs.

Cytoplasmic RNAs

The SRP may not be restricted only to eukaryotes but SRP receptor related molecules might be present in every organism (Lütcke 1995). How central conserved motifs in regulatory RNAs are is illustrated by complementation experiments between 7SL RNA from eukaryotes and the prokaryotic 4.5S RNA. Ribes et al. (1990) showed successful complementation after gene disruption of the essential 4.5S *E.coli* RNA by a plasmid supplying 7SL RNA in *trans*. The small 7SK RNA has been shown to increase strongly upon oncogenic transformation, its secondary structure may perhaps participate in transformation-dependent *c-myc* deregulation, which would be a nuclear event (Luo et al. 1997). Though several other cytoplasmic RNAs are known and partly studied (Table 2.2), the roles and functional motifs of cytoplasmic RNAs are not well characterized. This is an obvious area, which would profit from further research.

Xist (x-chromosome inactivation) RNA

While males possess an X and Y chromosome in mammals, females possess two X chromosomes. To allow equal dosage of the genes on the X chromosome between the two sexes to occur, one of the female X chromosomes can be inactivated. Inactivation of the chromosome can be visualized as a Barr body (Barr et al. 1949), in which the chromosome takes a compact heterochromatin form. Other organisms solve the problem of dosage compensation by different means, for example in *Drosophila*, the single male X chromosome is up-regulated and expresses twice the amount of gene products as its female counterpart. In nematodes the opposite occurs and the two female X chromosomes down-regulate their gene expression (Hendrich and Willard 1995).

The inactivation of the X-chromosome in mammals lies under the control of a *cis* acting locus termed the X inactivation centre (XIC) (Clemson et al. 1996; Penny et al. 1996). A putative canditate gene implicated in the control of X-inactivation is the *XIST* gene (for X inactivation specific transcipt) (Brown et al. 1991; Lyon 1996). The *XIST* gene is transcribed into a very large 17 kb cDNA which contains no open reading frames and thus appears to be non-translated (Brockdorff et al. 1992; Brown et al. 1992). This gene is exclusively expressed by the inactive but not by the active chromosome (Rastan 1983; Rastan and Brown 1990; Brown et al. 1992; Clemson et al. 1996, Lee et al. 1996). The gene product is intimately associated with the inactive X chromosome. It acts in *cis* with the X chromosome and inactivates it upon binding (Clemson et al. 1996). The mode of the inactivation is as yet poorly understood but it is thought that upon interaction with the *XIST* RNA the chromatin on the X chromosome is remodeled (Herzing et al. 1997; Lee and Jaenisch 1997; Willard and Salz 1997). The mode of inactivation seems to be mediated by RNA stabilization. On the mouse *XIST* gene it has been shown that alternate promoter usage gives rise to distinct stable and unstable RNA isoforms. Unstable *XIST* transcript initiates at an upstream promoter, whereas stable *XIST* RNA is transcribed from promoters further downstream. Stable *XIST* RNA accumulates on the inactivated X chromosome (Johnston et al. 1998). Ectopic expression of *XIST* in murine embryonic stem cells leads to inactivation of the X chromosome. This exhibits the associated hallmark phenotypes, i.e. heterochromatin formation, hypoacetylation of histone H4, delay in replication and formation of the RNA-Barr-body complex (Herzing et al. 1997; Lee and Jaenisch 1997), confirming the role of the *XIST* RNA in X chromosome inactivation. Penny et al. (1996) have shown that X inactivation requires the transcription of the *XIST* gene, as a deletion of 7 kb in the first coding region of *XIST* in ES cells leads to the failure to inactivate the X chromosome.

In *Drosophila*, sex determination relies on the X chromosome to autosome ratio. In males dosage compensation is achieved through hyper-transcription of the male X chromosome. Four specific *msl* (male specific lethal) genes have been identified which are involved in the dosage compensation mechanism (reviewed in Kuroda et al. 1993; Gebauer et al.; 1997). Hypertranscription is associated with an alteration of chromatin structure which involves a msl protein complex which "paints" the X chromosome (Kelley and Kuroda 1995).

Two other genes, *roX1* and *roX2* (*roX* for RNA on the X chromosome), have been identified by Meller et al. (1997) and Amrein and Axel (1997), which, akin to the *XIST* gene product, do not encode for a protein. Both RNAs are produced in a male-specific manner and expressed preferentially in the CNS. The role of the *roX1* and 2 genes is opposite of the *XIST* RNA, they change the X chromosome structure such that it is activated and lead to hypertranscription.

The number of X chromosomes must be correctly assessed via a "counting mechanism" before an X chromosome can be inactivated. A report by Nicoll et al. (1997) has shown how this is achieved in *C. elegans*. Males carry only one X chromosome, while hermaphrodites carry two. Dosage compensation involves the gene *xol-1* which, when active, leads to the male-specific fate and, when inactive, leads to the hermaphrodite fate. The X chromosome contains specific counting signals (Akerib and Meyer 1994), which affect the activity of xol-1. Two different mechanisms act in concert to repress the activity of *xol-1* in XX hermaphrodites. One involves the repression of the *xol-1* transcription, whereas the other employs a post-transcriptional strategy in which a putative RNA-binding protein termed fox-1 binds to *xol-1* and further decreases its activity.

Imprinting of X in mammals is controlled by the antisense XIST gene, TSIX (Lee 2000). TSIX is maternally expressed and mice carrying a TSIX deletion show normal paternal but impaired maternal transmission. Maternal inheritance occurs infrequently, with surviving progeny showing intrauterine growth retardation and reduced fertility. Transmission ratio distortion results from disrupted imprinting and post-implantation loss of mutant embryos. In contrast to effects in embryonic stem cells, deleting TSIX causes ectopic X inactivation in early male embryos and inactivation of both X chromosomes in female embryos, indicating that X chromosome counting cannot override TSIX imprinting. These results highlight differences between imprinted and random X inactivation but show that TSIX regulates both. This suggests that an imprinting center is located within TSIX.

H19 RNA and Imprinting

The *H19* gene was originally identified as an abundant mRNA under the control of a *trans*-acting locus termed *raf* (Pachnis et al. 1984). The 2.5 kb RNA lacks an ORF and does not encode a protein (Brannan et al. 1990). Expression of the gene in transgenic mice results in prenatal lethality, indicating that the gene product is tightly controlled in a dose dependent manner (Brunkow and Tilghman 1991). *H19*, located on chromosome 7 in mice, is expressed exclusively from the maternal allele in a sequence, which also carries additional imprinted genes (Bartolomei et al. 1991). Genomic imprinting is a feature of certain genes which are marked as being derived either from the maternal or paternal lineage and allowing expression from either the maternal or paternal lineage. It is thought that DNA methylation may represent a molecular "marker" which tags a given gene and thus regulates its expression (Li et al. 1993). Three different genes, i.e *H19*, *Igf-2r* and *Igf-2*, show differential expression patterns depending whether they are derived from maternal or paternal origin (Sasaki et al. 1992; Bartolomei et al. 1993; Brandeis et al. 1993; Ferguson-Smith et al. 1993; Stöger et al. 1993). Upon examination of the expression patterns of the three imprinted genes in mutant transgenic mice, which lack DNA methyltransferase activity, it was shown that the imprints of all three genes were lost. The normally repressed *H19* gene was activated and the active *Igf-2* and *Igf-2r* genes were repressed leading to the premature death of the mouse embryos (Li et al. 1993). It is thought that the allele-specific modification patterns may be mediated by impriting centres, which would contain *cis*-acting elements, marked differentially in maternal and paternal gametes (Eden and Cedar 1995). Such an imprinting center has been found on the human chromosome 15, which is responsible for a neurobehavioral disorder termed Prader-Willi-syndrome (PWS) (Wevrick et al. 1994). This leads to changes in the methylation and expression pattern of other imprinted genes (Sutcliffe *et al.* 1994; Buiting et al. 1995). The imprinted gene in the PWS region also encodes a non-translatable RNA molecule akin to *H19* and *XIST* (Wevrick et al. 1994), although it is unclear how these RNAs might function on the imprinting center. Deletion of a silencer element in *igf2* results in loss of imprinting independent of *H19* (Constancia et al. 2000). It has been suggested that *H19* RNA may exhibit tumor supressor activity (Hao et al. 1993).

Caenorhabditis elegans lin-4 and let-7 RNA

C. elegans undergoes stage-specific changes which are temporally synchronized. Several such genes are responsible for the control of larval development, which encompasses the stages L1 to L4 (Ambros and Moss 1994; Ambros and Horvitz 1984). Temporal development is dependent on the *lin-14* gene, which encodes a nuclear protein. *Lin-14* protein levels are higher during the L1 stage and decline in L2, where they are barely detectable (Ruvkun and Giusto 1989; Wightman et al. 1993). Although *lin-14* is fully repressed in the late developmental stages its mRNA is still present, indicating that down-regulation occurs at a post-transcriptional level (Wightman et al. 1991, 1993). It has been shown that *lin-14* is down-regulated by the activity of *lin-4*, which encodes for a small RNA molecule. *Lin-4* is produced in two transcripts of 22 and 61 nts in length (Lee et al. 1993); of which the major 22 nts RNA exhibits sequence complementarity to seven regions of the 3' UTR of the *lin-14* mRNA. Binding of the *lin-4* RNA to these regions is necessary and sufficient for repression of *lin-14* (Wightman et al. 1993). Single base pair changes, which disrupt the interaction, also abolish *lin-4* function (Lee et al. 1993).

The RNA-RNA interaction is formed via *lin-4/lin-14* antisense sequences and involves the common sequence ACUCC of *lin-4* RNA, which is located in a loop region (Fig. 2.18). A bulged C residue appears to be crucial, as mutations, which substitute the C to U abolishes *lin-4* function (Wightman et al. 1993). It is thought that this might be recognized by a *trans*-acting factor (Ha et al. 1996). Additional sequences found outside the repeats are also thought to be important for the repression mechanism, as genes which carry all the seven repeats, but lack conserved 3' UTR portions exhibit a loss of repression in varying degrees (Wightman et al. 1993). These additional sequences might be important for mediating proper folding of the RNA to promote sense/antisense RNA interactions. How does the repression come about?

The C. *elegans* heterochronic gene-pathway that consists of a cascade of regulatory genes which are temporally controlled to specify the timing of developmental events. *Let-7* is such a heterochronic switch gene. Loss of it causes reiteration of larval cell fates during the adult stage, whereas increased *let-7* gene dosage causes precocious expression of adult fates during larval stages. *Let-7* encodes a temporally regulated 21 nts RNA which is complementary to elements in the 3' UTR of the heterochronic genes *lin-14, lin-28, lin-41, lin-42* and *daf-12*. This indicates that their expression may be directly controlled by *let-7*. A reporter gene bearing the *lin-41* 3' UTR is temporally regulated in a *let-7*-dependent manner. *lin-4* does not only negatively regulate *lin-14*, but as well *lin-28* through RNA-RNA interaction. Reinhart et al. (2000) propose that the sequential stage-specific expression of the *lin-4* and *let-7* regulatory RNAs triggers transitions in the complement of heterochronic regulatory proteins to coordinate developmental timing.

Lin-4 RNA does not seem to control the stability of the *lin-14* mRNA as this is present in all stages of development (Wightman et al. 1993). It has been suggested that the *lin-4* RNA might mask the *lin-14* mRNA such that it is not accessible for the translational machinery (Wickens and Takayama 1994); or it might lead to its mislocalization where it cannot be translated (Lee et al. 1993).

Lin-4 RNA has been shown to bind the 3' UTR of *lin-28* in a fashion similar to that in *lin-14*. *Lin-28* codes for a cold-shock domain protein, which controls developmental timing. Binding occurs to a single *lin-4* complementary site in the 3' UTR causing down-regulation of *lin-28* (Moss et al. 1997).

Telomerases: Maintaining the Ends of the Chromosomes

DNA replication is discontinuous. DNA polymerase is a unidirectional enzyme, synthesizing from 5' to 3' direction. A conventional DNA polymerase replicates the leading strand, the lagging opposite strand requires a different mechanism for its 5' to 3' replication. Perhaps as a reminiscence of the postulated RNA world (reviewed in Gesteland et al. 1999) short 8-12 bp RNA primers are synthesized at intervals. Conventional DNA polymerase fills in the gaps between them. The RNA primers are removed afterwards (so DNA replaces RNA as it has done probably also during phylogeny) and the 8-12 bp gaps are "repaired". However, at the end of the discontinuous strand (in contrast to the circular chromosomes in bacteria) there is a small gap left. This poses the problem that the chromosomes would be shortened after each replication cycle. To counteract this, most eukaryotes maintain the length of the ends by employing a novel DNA polymerase, termed the telomerase (Greider and Blackburn 1985, 1987; Blackburn and Greider 1995).

Telomerase is a ribonucleoprotein enzyme capable of extending the 3' end of the chromosome. The RNA component (8-30 nts in length) has been cloned from a number of organisms including yeast (Singer and Gottschling 1994) and man (Feng et al. 1995). It is central to provide the missing template for extension (Fig. 6.4). It base-pairs with the single-stranded 3' end of the chromosomal DNA, it extends, protects and caps chromosmal ends with telomerase repeats by repeated translocations of the telomerase complex (Zakian 1995; Fig. 6.4). By the conventional DNA polymerase a complementary strand is synthesized leading again to a short gap at the end after removal of the RNA primer.

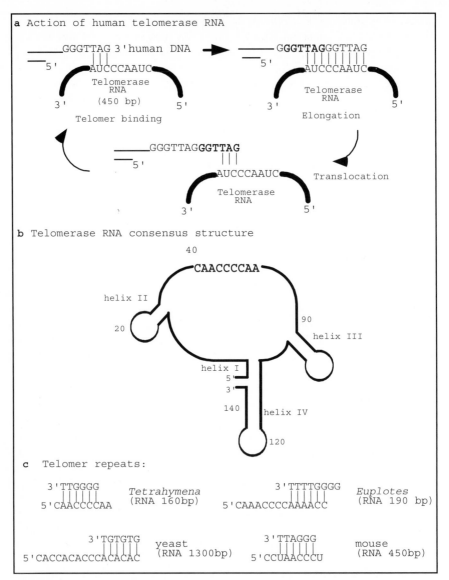

Fig.6.4a-c. Telomerase RNA. **a** Schematic cycle showing the interaction between human telomerase RNA and human chromosomal DNA. One chromosomal DNA Telomer repeat is always highlighted *in bold;* the cycle is repeated several times till the required length is achieved. **b** Tetrahymena telomerase RNA consensus structure (see McCormick-Graham and Romero 1995 for species examples). The templating domain is shown *in bold.* **c** Comparison of different telomerase RNA templating sequences and telomers (Autexier and Greider 1996). The length of the RNA is given *in brackets*, base pairing to DNA as well as orientation of DNA (*top*) and RNA strand (telomerase) is indicated

Using a phylogenetic comparative analysis on 32 telomerase genes from a variety of vertebrates. Chen et al. 2000 propose a secondary structure of telomerase RNA containing four structural domains conserved in all vertebrates. Ten helical regions of the RNA are universally conserved while other regions vary in length and sequence. The proposed vertebrate telomerase RNA is remarkably similar and evolutionarily conserved compared to ciliate RNA telomerase. Several different RNA motifs serve as templates for the telomerase repeats (Fig. 6.4; Greider and Blackburn 1989; Blackburn and Greider 1995). The complete telomerase RNA has a specific secondary structure (Fig. 6.4). All eukaroytic telomers which have been analyzed possess a stretch of G-rich repeats that run in the 5' to 3' direction. Telomere sequences vary considerably across species (Fig. 6.4), e.g Telomerase in *Tetrahymena* adds the repeat TTGGGG (Blackburn and Gall 1978), whereas in *Kluyveromyces lactis* it is ACGGATTTGATTAGGTAT GTGGTGT (McEachern and Blackburn 1994). Telomere shortening is associated with cellular ageing and senescence and limits somatic cell division potential (Harley and Villeponteau 1995). Immortalized cells, germ cells and tumours show a high level of telomerase activity (de Lange 1994; Kim et al. 1994). Telomerase inhibitors may be potential anti-cancer therapeutics.

Malfunction of telomerase causes the X-linked form of dyskeratosis congenital (DKC). Dyskerin is a putative pseudouridin synthase. It is associated not only with defects in rRNA processing and snoRNAs but also with defects in telomerase RNA due to a shared motif (H/ACA). DKC cells have a lower level of telomerase activity and its RNA. Limited proliferative capacity results in defects in skin and bone marrow (Mitchell et al. 1999). Telomerase RNA mutations lead to incorporation of this altered sequence into the telomere DNA *in vivo* (Singer and Gottschling 1994) and *in vitro* (Feng et al. 1995). The protein telomerase component cloned from *S.pombe* and man are reverse transcriptases, but clearly different from retroviral and retrotransposons (Nakamura et al. 1997).

Research on the mechanistic aspects of telomere synthesis by yeast telomerase showed that a conserved RNA structure adjacent to the template rather than the sequence specifies the template boundary. Disruption of the structure caused DNA synthesis to proceed beyond normal template boundary, resulting in altered telomere sequences, telomere shortening, and cellular growth effects, revealing a specific function for an RNA structure in the enzymatic action of telomerase (Tzfati et al. 2000).

Mammalian telomerase RNAs are two- to threefold bigger than ciliate counterparts, an extra domain resembling a box H/ACA snoRNA is present. The human telomerase (hTR) RNA H/ACA domain is essential *in vivo* for accumulating hTR as well as for its 3' end processing and for telomerase activity. Maturation of full-length hTR and its assembly occurs from an mRNA promoter-driven RNA polymerase II transcript. Furthermore, a small percentage of hTR is associated with nucleoli (Mitchell et al. 1999).

Telomere length is monitored (Cooper et al. 1997; Marcand et al. 1997; Van Steensel and de Lange 1997) by counting proteins bound to the telomers (human

TRF 1, *S. pombe* Taz1p and yeast Rap1, which all share a common Myb-like DNA binding domain). Extension (Fig. 6.4) is shut off when the required length is achieved.

Alternative Splicing

Alternative splicing of the mRNA from the gene NADPH oxidase homologue 1 (NOH-1) leads to a 191 amino acid protein with properties indistinguishable from native H^+-channels for whom a molecular identification has not yet been achieved (Banfi B et al. 2000). Earlier, Iverson et al. (1997) showed that differential splicing of *Drosophila* Shaker (*Sh*) gene transcripts regulates the tissue-specific expression of kinetically distinct potassium ion channels throughout development. Jones et al. (1999) revealed the role of alternative splicing of Ca^{2+} -activated K^+ channels in auditory hair cells. A molecular mechanism for hair cell frequency tuning involving differential expression of different K^+ channel alpha-subunits in conjunction with an expression gradient of the regulatory beta-subunit is suggested.

Small RNAs in E.coli

E.coli cells downregulate the expression of outer membrane transport protein F (*ompF* gene; Andersen et al. 1989) by a 93 nts MicF RNA which base pairs to the 5' UTR of the *ompF* transcript, thereby either destabilizing and/or masking it from the translational machinery (Wagner and Simons 1994). Altuvia et al. (1997) identified a novel 109 nts, abundant and stable RNA, termed oxyS. OxyS RNA has three stem-loop structures (Fig. 6.5) and is induced upon oxidative stress and dramatically changes the expression of several genes by either activating or repressing them, acting as an anti-mutator.

```
                        C U
                    U       C
                    U       A
                     U A
                    G C
        5'          A U
         G C G A G C        A A U A A C U A A A G C C A A C G U G A A C U U U U   3'
         A U              G C                                              C G
         A U                        7 0                                    G C
         A U                                                               G C
         C G                                                               A U
         G C                                                               U A
         G C                                                               C G
       A   40                                                            U   G
         G C                                                           C   A
         C G                                                            C   100
         G U
         G C
       C   C
         C   A
         A U
         C G
       C   A
        U A
         C G
        U   U
      U       U
      U         C
   20 U             C
        A   A   C
            A
```

6.5. oxyS RNA. oxyS RNA is induced during the stress response in *E.coli*. The secondary structure drawn is according to standard folding predictions and primary sequence. Positons of nucleotides are indicated. (Altuvia et al. 1997)

7. Outlook

7.1 New Types and Classes of Regulatory RNAs Come up at an Unprecedented Pace

Where may new regulatory elements and RNA motifs be identified in the future? Though the concrete areas may be difficult to predict, genomics is changing everything. We are experiencing the advent of full genomic sequences for nearly 50 prokaryotic genomes and for about further 150 microbial genomes, the sequencing is in progress; many of these will be finished within the next 2 years. Complete eukaryotic genomes from *C.elegans*, *A.thaliana*, *Drosophila* and even man are almost finished - even if in the rush among competing scientists and consortia to be the first, these eucaryotic genome sequences will still require numerous corrections and updates (detailed genome sequencing lists and literature references, both continously updated, are available at http://www.tigr.org/tdb). This stock of new genome data will doubtless dramatically increase the number of potential candidate structures as well as, after some time, confirmed regulatory RNA motifs and elements.

However, even now new types of RNA structures keep coming up with an increasing speed, besides new telomerases and previously unexpected large-scale RNA structures such as X inactivation by XIST RNA and its own inactivating antisense RNA TSIX (Lee 2000), these are for example regulatory RNAs in development such as let-7 Reinhart et al. (2000) and a host of new snoRNAs in Archaea (Gaspin et al. 2000). However, also the levels on which RNA exerts function increase, for example RNAs involved in RNA silencing and RNA degradation (Dalmay et al. 2000), as well as RNA interference (Mourrain et al. 2000). The latter citation stresses that many more RNA interference examples should be present in the *Arabidhopsis thaliana* genome waiting to be discovered; however, our bold claim would be that also in the other genomes not only many further RNAs of this type but also different other regulatory steps and levels are awaiting to be unveiled; The recent description of a new regulated pathway for nuclear pre-mRNA turnover (Bousquet-Antonelli et al. 2000) in yeast, incidentally another completely sequenced organism, underlines this claim.

7.2   Established Areas of Research

Regarding specific areas of research on well-established RNA motifs and regulatory elements, splicing remains a prosperous area of research. The dynamics of different interactions have now become a major focus (see Chap. 5) and the splicing process is studied and known now to a high dgreee of detail (see Chaps. 2 and 6). However, the overall metabolism of RNA in the nucleus should receive more attention and more imaginative research, a first hint of what still remains to be discovered are the novel RNA degradation pathways recently discovered (see above). Furthermore, accumulating data from large-scale gene expression analysis indicate that roughly 25% or sometimes even more of all transcripts in a typical eukaryotic nucleus come from either repetitive areas of the genome or transposons, or are processed RNA precursors and introns - and why the cell spends that much of its machinery and energy on this type of RNA turnover still awaits an interesting answer.

Aberrant reactions of the standard pathways from splicing and maturating mRNA are now found in many more instances. Thus *trans*-splicing was originally considered to be only an exceptional reaction in trypanosomes; however, new additional examples keep on being discovered in other organisms. Besides indications of *trans*-splicing reactions being more widespread and also, e.g. occurring in the immune system of higher organisms (Chap. 6) the list of RNA-editing examples is steadily increasing. This poses also the question of how many genes there are in the organism into a new context: for instance, there is always a chance that hidden somewhere in the genome there may be encoded a small guide RNA allowing, via RNA editing, a yield, for example, of two different receptors out of one receptor gene (see Chap. 6). This effect becomes quite interesting, as it may help to reconcile the surprising complexity of the human nervous system, requiring a large number of proteins with latest estimates for the number of human genes being surprisingly low. The discovery of ATAC introns in splicing shows that more variants of splicing are in stock. It further points to one approach to reveal further variants. A close examination of different introns and their splicing motifs should reveal further new interaction partners and splicing variants.

A major breakthrough in the investigation of ribosomal RNA and its motifs was the structural description of the large ribosomal subunit and that - as has long been suspected - there is really catalytical RNA critical for the active centre (Ban et al. 2000). Current research proves that many steps of the ribosomal RNA processing pathway are not yet understood in detail but that there is steady progress (Fig. 6.3). The RNA motifs involved, e.g. the different ribosomal RNA sites recognized, provide a tool to identify the interacting protein partners, e.g. by mutation of important cleavage sites. Compensatory mutations (Chap.s 3 and 6) delineate which RNAs interact directly and where.

Guide RNAs for methylation and pseudouridylation probably exist also for modified nucleotides in other RNA molecules such as tRNA, snRNAs and snoRNAs; this is a further area where we expect new RNA molecules to be discovered - and this includes all three kingdoms of life, notably Archaea (see above).

The active participation of rRNA in translation has been realized, but many details including involved RNA motifs, essential and non-essential regions for different steps still have to be described (Green and Noller, 1997).

The discovery of the 10Sa RNA as a tRNA-like structure (10Sa RNA is also called tmRNA) involved in correcting missing stop codons of mRNA (Chap. 2 and 5) shows that also here new RNA motifs and RNA pathways are being discovered.

Research on mRNA experienced a boost in new motifs in the 5´and 3´ UTR as well as new levels of regulation such as RNA interference and silencing. Our earlier claim (Dandekar and Sharma, 1998) that there should be many more examples for antisense RNAs also in higher cells has been well confirmed. Regulatory signals in the open reading frame of an mRNA are even more hidden, because there is the additional constraint of the reading frame. Nevertheless, known examples from viral and bacterial translation indicate how also in this position different signals are encoded in the RNA structure and influence translation. The examples so far revealed for eukaryotes are more rare, but this is, in our eyes, more a result of the more covered implementation (see Chap. 4) of these signals in the context of a reading frame. There may be a wealth of structure-function RNA interactions literally hidden, which should be further examined. Papers on the connection between RNA structure, translation and protein folding (Thanaraj and Argos 1996; Brunak and Engelbrecht 1996) give an example.

Engineered catalytic or regulatory RNA for biotechnological applications have still some way to go before they become a general method. The challenge of effective delivery *in vivo* and *in vitro* remains. However, we have to note that in those cases where such obstacles could be circumvented (e.g. Simons et al 1992) the antisense approach presents a very promising technology to influence gene expression. Virus-mediated expression of RNAs and alternative pharmacological approaches (e.g. liposomes) are other ways to tackle the obstacle of effective in vivo and in vitro administration, but they need further enhancements. Furthermore, stable chemical modifications (Table 5.2) allow, at least in vitro, far more applications than previously expected. Peptide or chemical synthesized mimics of the antisense RNA are another option, with the potential to provide chemical stability and efficient delivery at the same time.

7.3   Conclusion

Both the established and the new areas where RNA motifs and regulatory elements have been found to play an important functional role strongly indicate that the rapid expansion of this exciting field will last on. The availability of complete genomes and furthermore of gene expression and proteome data will introduce a new dimension to this game. The unexpectedly low number of human genes published in the draft sequences of the human genome (International Human Genome Sequencing Consortium, 2001) has further increased the general interest in all forms of RNA processing and alternative splicing of RNA. Besides their role in information transport, RNA molecules can be participating as actively in the cellular metabolism as are proteins - and it is both challenging and rewarding to find all the interesting cellular interactions where RNA motifs and regulatory elements play a decisive role.

8. Figure-Acknowledgement

We thank for the kind permission to reproduce several figures from other publications and in particular, we acknowledge here the following contributions (further material and references are given in the figure legends):

Fig. 2.8 Academic Press (25.10.00; authors Deckman and Draper, JMB 196, 323-332,Fig. 9)

Fig. 2.10b CSHL Press (11.12.00; authors Burge et al., The RNA World, 2nd Ed., p.527)

Fig. 2.19a CSHL Press (11.12.00; authors Pestova et al., Genes & Development 12, 67-83 (1998) Fig. 1b of p. 69)

Fig. 2.19b Elsevier Science (24.1.01; authors Wilson et al., Cell 102, p.512, (2000).

Fig. 2.21, 2.22 FASEB Journal (1.11.00; authors Long and Uhlenbeck; FASEB 7,25-30 (1993))

Fig. 2.25 Nature Publishing group (12.12.00; authors Ikawa et al., Nature structural biology 7, 1032 (2000); Fig. 1)

Fig 2.26 AAAS (7.12.00; authors Guo et al., Science 289, 453; Fig. 1b)

Fig. 5.1 MacMillan Magazines Ltd. (30.10.00; authors Sullenger and Cech, Nature 371, 619-622 (1994); Fig. 1)

9. References

Agalarov SC, Sridhar Prasad G, Funke PM, Stout CD, Williamson JR. Structure of the S15,S6,S18-rRNA complex: assembly of the 30S ribosome central domain. Science 2000; 288: 107-113.

Ahn AH, Kunkel LM. The structure and functional divergence of dystrophin. Nat Gen 1993; 3: 283-291.

Ahringer J, Kimble J. Control of the sperm-oocyte switch in *Caenorrhabiditis elegans* hermaphrodites by the fem-3 3' untranslated region. Nature 1991; 349: 346-348.

Akerib CC, Meyer BJ. Identification of X chromosome regions in *Caenorhabditis elegans* that contain sex-determination signal elements. Genetics 1994; 138: 1105-1125.

Allain FH, Gubser CC, Hower PW, Nagai K, Neuhaus D, Varani G. Specificity of ribonucleoprotein interaction determined by RNA folding during complex formation. Nature 1996; 380: 646-650.

Allain FH, Gilbert DE, Bouvet P, Feigon J. Solution structure of the two N-terminal RNA-binding domains of nucleolin and NMR study of the interaction with its RNA target. J Mol Biol 2000; 303: 227-241.

Altman S. Kirsebom L, Talbot S. Recent studies of ribonuclease P. FASEB J 1993; 7: 7-14.

Altona C. Classification of nucleic acid junctions. J Mol Biol 1996; 263: 568-581.

Altuvia S, Weinstein-Fischer D, Zhang A, Postow L, Storz G. A small, stable RNA-induced by oxidative stress: role as a pleiotropic regulator and antimutator. Cell 1997; 90: 43-53.

Amarasinghe GK, De Guzman RN, Turner RB, Summers MF. NMR structure of stem-loop SL2 of the HIV-1 psi RNA packaging signal reveals a novelvA-U-A base-triple platform. J Mol Biol 2000; 299: 145-156.

Ambros V, Horvitz HR. Heterochronic mutants of the nematode *Caenorhabditis elegans*. Science 1984; 226: 409-416.

Ambros V, Moss EG. Heterochronic genes and the temporal control of *C. elegans* development. Trends Genet 1994; 10: 123-127.

Amrein H, Axel R. Genes expressed in neurons of adult male *Drosophila*. Cell 1997; 88: 459-469.

Andersen J, Delihas N, Ikenaka K, Gren PJ, Pines O, Ilercil O, Inouye, M. The isolation and characterization of RNA coded by the micF gene in *Escherichia coli*. Nucl Acids Res 1987; 15: 2089-2101.

Andersen J, Forst SA, Zhao KJ, Inouye M, Delihas N. The function of micF RNA. micF RNA is a major factor in the thermal regulation of OmpF protein in *Escherichia coli*. J Biol Chem 1989; 264: 17961-17970.

Andersson U, Antonicka H, Houstek J, Cannon B. A novel principle for conferring selectivity to poly(A)-binding proteins: interdepence of two synthase beta-subunit mRNA-binding proteins. Biochem J 2000; 346: Pt 1:33-39.

Andresson T, Ruderman JV. The kinase Eg2 is a component of the *Xenopus* oocyte progesteron-activated signaling pathway. EMBO J. 1998; 17: 5627-5637.

Ansel-McKinney P, Gehrke L. RNA determinants of a specific RNA-coat protein peptide interaction in alfalfa mosaic virus: conservation of homologous features in ilarvirus RNAs. J Mol Biol 1998; 278: 767-785.

Antes T, Costandy H, Mahendran R, Spottswood M, Miller D. Insertional editing of mitochondrial tRNAs of *Physarum polycephalum* and *Didymium nigripes*. Mol Cell Biol 1998; 18: 7521-7527.

Ares M, Igel AH. Lethal and temperature-sensitive mutations and their suppressors identify an essential structural element in U2 small nuclear RNA. Genes Dev 1990; 4: 2132-2145.

Arn EA, Macdonald PM. Motors driving mRNA localization: new insights from in vivo imaging. Cell 1998; 95: 151-154.

Askari FK, McDonnell WM. Molecular medicine: antisense-oligonucleotide therapy. New Engl J Med 1996; 334: 316-318.

Atkin AL, Schenkman LR, Eastham M, Dahlseid JN, Lelivelt MJ, Culbertson MR. Relationship between yeast polyribosome and upf proteins required for nonsense mRNA decay. J Biol Chem 1997; 272: 22163-22172.

Autexier C, Greider CW. Telomerase and cancer: revisiting the telomere hypothesis. Trends Biochem Sci 1996; 21: 387-391.

Avni D, Biberman Y, Meyuhas O. The 5' terminal oligopyrimidine tract confers translational control on TOP mRNAs in a cell-type and sequence context-dependent manner. Nucl Acids Res 1997; 25: 995-1001.

Bachellerie J-P, Cavaillé J. Guiding ribose methylation of rRNA. Trends Bioch Sci 1997; 22: 257-261.

Bachellerie JP, Michot B, Nicoloso M. Antisense snoRNAs: a family of nucleolar RNAs with long complementarities to rRNA. Trends Biochem Sci 1995; 20: 261-264

Balakin AG, Smith L, Fournier MJ. The RNA world of the nucleolus: two major families of small RNAs defined by different box elements with related functions. Cell 1996; 86: 823-834.

Ban N, Nissen P, Hansen J, Moore PB, Steitz TA. The complete atomic structure of the ribosomal subunit at 2.4 Å resolution. Science 2000; 289: 905-920.

Baron C, Boeck A. The length of the aminoacyl-acceptor stem of the selenocystein- specific tRNA^{Ser} of *Escherichia coli* is the determinant for binding to elongation factor SELB or Tu. J Biol Chem 1991; 266. 20375-20379.

Baron C, Heider J, Boeck A. Interaction of translation factor SELB with the formate dehydrogenase H selenopolypeptide mRNA. PNAS 1993, 90. 4181-4185.

Barr ML, Bertram EG. A morphological distinction between neurones of the male and female, and the behaviour of the nucleolar sattelite during accelerated nucleoprotein synthesis. Nature 1949; 163; 676-677.

Barth C, Greferath U, Kotsifas M, Fisher PR. Polycistronic transcription and editing of the mitochondrial small subunit (SSU) ribosomal RNA in *Dictyostelium discoideum*. Curr Genet 1999; 36: 55-61.

Bartolomei MS, Webber AL, Brunkow ME, Tilghman SM. Epigenetic mechanisms underlying the imprinting of the mouse H19 gene. Genes Dev 1993; 7: 1663-1673.

Bartolomei MS, Zemel S, Tilghman SM. Parental imprinting of the mouse H19 gene. Nature 1991; 351: 153-155.

Basillo C, Wahba AJ, Lengyel P, Speyer JF, Ochoa S. Synthetic polynucleotides and amino acid code. PNAS 1962; 48: 613-616.

Bass BL. Physarum-C the difference. Nature 1991; 349: 370-371.

Bass BL. Splicing: the new edition. Nature 1991; 352: 283-284

Bass BL. An I for editing. Curr Biol 1995; 5: 598-600.

Baum CL, Teng BB, Davidson NO. Apolipoprotein B messenger RNA editing in the rat liver. Modulation by fasting and refeeding a high carbohydrate diet. J Biol Chem 1990; 265: 19263-19270.

Bektesh S, Van Doren K, Hirsch D. Presence of the *Caenorhabditis elegans* spliced leader on different mRNAs and in different genera of neamtodes. Genes Dev 1988; 2: 1277-1283.

Belfort M. Prokaryotic introns and inteins: a panoply of form and function. J Bacteriol 1995; 177: 3897-3903.

Belostotsky DA, Meagher RB. Different organ specifc expression of three poly(A) binding protein genes from *Arabidopsis thaliana*. PNAS 1993; 90:6686-6696.

Belsham GJ, Jackson RJ. Translation initiation on picornavirus RNA. In: Sonenberg N, Hershey J, Mathews M (eds). Translational Control of Gene Expression, CSHL Press, 2000; 860-900.

Beltrame M, Henry Y, Tollervey D. Mutational analysis of an essential binding site for the U3 snoRNA in the 5' external transcribed spacer of yeast pre-rRNA. Nucleic Acids Res 1994; 22: 5139-5147.

Beltrame M, Tollervey D. Identification and functional analysis of two U3 binding sites on yeast pre-ribosomal RNA. EMBO J 1992; 11: 1531-1542.

Beltrame M, Tollervey D. Base pairing between U3 and the pre-ribosomal RNA is required for 18S rRNA synthesis. EMBO J 1995; 14: 4350-4356.

Benne R. RNA editing: is there a message? Trends Genet 1990; 6: 177-181

Benne R. RNA editing. Ellis Horwood, 1993.

Benne R. The long and short of it. Nature 1996; 380: 391-392.

Benne R, Van den Burg J, Brakenhoff JP, Sloof P, Van Boom JH and Tromp MC. Major transcript of the frameshifted coxII gene from trypanosome mitochondria contains four nucleotides that are not encoded in the DNA. Cell 1986; 46: 819-826.

Berget SM, Moore C, Sharp PA. Spliced segments at the 5' terminus of adenovirus 2 late mRNA. PNAS 1977; 74: 3171-3175.

Berglund JA, Chua K, Abovich N, Reed R, Rosbash M. The splicing factor BBP interacts specifically with the pre-mRNA branchpoint sequence UACUAAC. Cell 1997; 89: 781-787.

Bergmann KF, Gerin JL. Antigens of hepatitis delta virus in the liver and serum of humans and animals. J Infect Dis 1986; 154: 702-706.

Bernus I, Mitchell AM, Manley SW, Mortimer RH. Lack of membrane transport of l-thyroxine sulphate in the human choriocarcinoma cell line, JAr. Placenta. 2000; 21: 283-285.

Berry MJ. Recoding UGA as selenocysteine. In: Sonenberg N, Hershey J, Mathews M (eds). Translational Control of Gene Expression, CSHL Press 2000; 763-783.

Berry MJ, Banu L, Chen Y, Mandel SJ, Kieffer JD, Harney JW, Larsen PR. Recognition of UGA as a selenocysteine codon in Type I deiodiase requires sequences in the 3' untranslated region. Nature 1991; 353: 273-276.

Beyer K, Dandekar T, Keller W. RNA-ligands selected by cleavage stimulation factor (CstF) contain distinct sequence motifs that function as downstream elements in 3'-end processing or pre-mRNA. J Biol Chem 1997; *in press.*

Bhat GJ, Koslowsky DJ, Feagin JE, Smiley BL, Stuart K. An extensively edited mitochondrial transcript in kinetoplastids encodes a protein homologous to ATPase subunit 6. Cell 1990; 61: 885-894.

Bhattacharyya A and Blackburn EH. A functional telomerase RNA swap in vivo reveals the importance of nontemplate RNA domains. PNAS 1997; 94: 2823-2827.

Binder S, Marchfelder A, Brennicke A, Wissinger B. RNA editing in trans-splicing intron sequences of nad2 mRNAs in *Oenothera* mitochondria. J Biol Chem 1992; 267: 7615-7623.

Binder S, Marchfelder A, Brennicke A. RNA editing of tRNA(Phe) and tRNA(Cys) in mitochondria of Oenothera berteriana is initiated in precursor molecules. Mol Gen Genet 1994; 244: 67-74.

Black AC, Ruland CT, Luo J, Bakker A, Fraser JK, Rosenblatt JD. Binding of nuclear proteins to HTLV-II *cis*-acting repressive sequence (CRS) RNA correlates with CRS function. Virology 1994; 200: 29-41.

Black DL, Chabot B, Steitz JA. U2 as well as U1 small nuclear ribonucleoproteins are involved in pre-messenger RNA splicing. Cell 1985; 42: 737-750.

Blackburn EH . Telomerase. In: Gesteland RF, Cech TR, Atkins JF (eds) The RNA World., CSHL Press 1999; 609-635.

Blackburn EH. The end of the (DNA) line. Nat Struct Biol 2000; 7: 847-850.

Blackburn EH, Gall JG. A tandemly repeated sequence at the termini of the extrachromosomal ribosomal RNA genes in *Tetrahymena*. J Mol Biol 1978; 120:33-53.

Blackburn EH, Greider CW. Telomers. Plainview, CSHL Press 1995.

Blasco MA, Funk W, Villeponteau B, Greider CW.   Functional characterization and developmental regulation of mouse telomerase RNA. Science 1995; 269: 1267-1270.

Blum B, Bakalara N, Simpson L. A model for RNA editing in kinetoplastid mitochondria: "guide" RNA molecules transcribed from maxicircle DNA provide the edited information. Cell 1990; 60: 189-198.

Blum B, Sturm NR, Simpson AM, Simpson L. Chimeric gRNA-mRNA molecules with oligo(U) tails covalently linked at sites of RNA editing suggest that U addition occurs by transesterification. Cell 1991; 65: 543-550.

Blumenthal T. Trans-splicing and polycistronic transcription in *Caenorhabditis elegans*. Trends Genet 1995; 11: 132-136.

Boelens WC, Jansen EJR, van Venrooij WJ, Stripecke R, Mattaj IW, Gunderson SI. The human U1 snRNP-specific U1A protein inhibits polyadenylation of its own pre-mRNA. Cell 1993; 72:881-892.

Bohnen L. *Trans*-splicing of pre-mRNA in plants, animals and protists. FASEB J 1993; 7: 40-46.

Bonino F, Heermann KH, Rizzetto M, Gerlich WH. Hepatitis delta virus: protein composition of delta antigen and its hepatitis B virus-derived envelope. J Virol 1986; 58: 945-950.

Bork P, Gibson TJ. Applying motif and profile searches. Methods Enzymol 1996; 266:162-184.

Borner GV, Morl M, Janke A Paabo S. RNA editing changes the identity of a mitochondrial tRNA in marsupials. Embo J 1996; 15:5949-5957.

Boros I, Posfai G, Venetianer P. High copy number derivatives of the plasmid cloning vector pBR322. Gene 1984; 30:257-260.

Böstrom K, Garcia Z, Poksay KS, Johnson DF, Lusis AJ, Innerarity TL. Apolipoprotein B mRNA editing. Direct determination of the edited base and occurrence in non-apolipoprotein B-producing cell lines. J Biol Chem 1990; 265: 22446-22452.

Bosquet-Antonelli C, Henry Y, Gélugne JP, Caizergues-Ferrer M, Kiss T. A small nucleolar RNP protein is required for pseudouridylation of eukaryotic ribosomal RNAs. EMBO J 1997; *in press*

Boudvillain M, de Lencastre A, Pyle AM. A tertiary interaction that links active-site domains to the 5' splice site of a group II intron. Nature 2000; 406:315-318.

Bousquet-Antonelli C, Presutti C, Tollervey D. Identification of a regulated pathway for nuclear pre-mRNA turnover. Cell 2000; 102:765-775.

Bouthinon D, Soldano H. A new method to predict the consensus secondary structure of a set of unaligned RNA sequences. Bioinformatics 1999; 15: 785-798.

Brachet J. Reminiscences about nucleic acid cytochemistry and biochemistry. TIBS 1987; 12: 244-246.

Brand S, Bourbon HM. The developmentally regulated *Drosophila* gene rox8 encodes an RRM-type RNA-binding protein structure related to human TIA-1-type nucleolysins. Nucl Acids Res 1993; 21: 3699-3704.

Brandeis M, Kafri T, Ariel M, Chaillet JR, McCarrey J, Razin A, Cedar H. The ontogeny of allele-specific methylation associated with imprinted genes in the mouse. EMBO J 1993; 12: 3669-3677.

Brannan CI, Dees EC, Ingram RS, Tilghman SM. The product of the H19 gene may function as an RNA. Mol Cell Biol 1990; 10: 28-36.

Braun RE. Temporal control of protein synthesis during spermatogenesis. Int J Androl 2000; 23 Suppl 2:92-94.

Brehm K, Jensen K, Frosch M. mRNA Trans-splicing in the Human Parasitic Cestode *Echinococcus multilocularis*. J Biol Chem 2000; 275: 38311-38318.

Breitwieser W, Markussen FH, Horstmann H, Ephrussi A. *Oskar* protein interaction with Vasa represents an essential step in polar granule assembly. Genes Dev 1996; 10: 2179-2188.

Brenner S, Jacob F, Meselson M. An unstable intermediate carrying information from genes to ribosomes for protein synthesis. Nature 1961; 190:576-581.

Brockdorff N, Ashworth A, Kay GF, McCabe VM, Norris DP, Cooper PJ, Swift S, Rastan S. The product of the mouse XIST gene is a 15 kb inactive X-specific transcript containing no conserved ORF and is located in the nucleus. Cell 1992; 71: 515-526.

Brow DA, Guthrie C. Spliceosomal RNA U6 is remarkably conserved from yeast to mammals. Nature 1988; 334: 213-218.

Brown CJ, Ballabio A, Rupert JL, Lafreniere RG, Grompe M, Tonlorenzi R, Willard HF. A gene from the region of the human X-inactivation centre is expressed exclusively from the inactive X chromosome. Nature 1991; 349: 38-44.

Brown BD, Goldstein. Teaching old dogmas new tricks. Nature 1987; 330: 113-114.

Brown CJ, Hendrich BD, Rupert JL, Lafreniere RG, Xing Y, Lawerence J, Willard H. The human XIST gene: Analysis of a 17kb inactive X-specific RNA that contains conserved repeats and is highly localized within the nucleus. Cell 1992; 71: 527-542.

Brown CY, Mize GJ, Pineda M, George DL and Morris DR. Role of two upstream open reading frames in the translational control of oncogene mdm2. Oncogene 1999; 18: 5631-5637.

Brown BD, Zipkin ID, Harland RM. Sequence-specific endonucleolytic cleavage and protection of mRNA in *Xenopus* and *Drosophila*. Genes Dev 1993; 7: 1620-1631.

Brunak S, Engelbrecht J. Protein structure and the sequential structure of mRNA: alpha-helix and beta-sheet signals at the nucleotide level. Proteins 1996; 25:237-252.

Brunkow M, Tilghman SM. Ectopic expression of the H19 gene in mice causes prenatal lethality. Genes Dev 1991; 5: 1092-1101.

Bruzik JP, Maniatis T. Spliced leader RNAs from lower eukaryotes are trans-spliced in mammalian cells. Nature 1992; 360: 692-695.

Bruzik JP, Steitz JA. Spliced leader RNA sequences can substitute for the essential 5' end of U1 RNA during splicing in a mammalian *in vitro* system. Cell 1990; 62: 889-899

Buettner C, Harney JW, Berry MJ. The Caenorhabditis elegans homologue of thioredoxin reductase contains a selenocysteine insertion sequence (SECIS) element that differs from mammalian SECIS elements but directs selenocysteine incorporation. J Biol Chem 1999; 274: 21598-21602.

Buiting K, Saitoh S, Gross S, Dittrich B, Schwartz S, Nicholls R, Horsthemke B. Inherited microdeletions in the Angelman and Prader-Willi syndromes define an imprinting centre on human chromosome 15. Nat Genet 1995; 9: 395-400.

Burge CB, Tuschl T, Sharp PA . Splicing of precursors to mRNAs by the spliceosome. In: Gesteland RF, Cech TR, Atkins JF (eds) The RNA World, CSHL Press 1999; 525-560.

Burke DH, Gold L. RNA aptamers to the adenosine moiety of S-adenosyl methionine: structural inferences from variations on a theme and the reproducibility of SELEX. Nucleic Acids Res 1997; 25:2020-2024.

Burtis KC, Baker BS. *Drosophila* doublesex gene controls somatic sexual differentiation by producing alternatively spliced mRNAs encoding related sex-specific polypeptides. Cell 1989; 56, 997-1010.

Caffarelli E, Fragapane P, Ghering C, Bozzoni I. THe accunmulation of mature RNA for *the Xenopus laevis* ribosomal protein L1 is controlled at the level of splicing and turnover of the precursor RNA. EMBO J 1987; 6: 3492-3498.

Camaschella C, Zecchina G, Lockitch G, Roetto A, Campanella A, Arosio P, Levi S. A new mutation (G51C) in the iron-responsive element (IRE) of L-ferritin associated with hyperferritinaemia-cataract sindrome decreases the binding affinity of the mutated IRE for iron-regulatory proteins. Br J Haematol 2000; 108: 480-482.

Canete-Soler R, Schlaepfer WW. Similar poly(C)-sensitive RNA-binding complexes regulate the stability of the heavy and light neurofilament mRNAs. Brain Res 2000; 867: 265-279.

Caponigro G, Muhlrad D, Parker R. A small segment of the MAT alpha 1 transcript promotes mRNA decay in *Saccharomyces cerevisiae*: a stimulatory role of rare codons. Mol Cell Biol 1993; 13: 5141-5148.

Caponigro G, Parker R. mRNA turnover in yeast promoted by the MATalpha1 instability element. Nucleic Acid Res 1996; 24: 4304-4312.

Carballo E, Lai WS, Blackshear PJ. Feedback inhibition of macrophage tumor necrosis factor-alpha production by tristetraprolin. Science 1998; 281: 1001-1005.

Carrara G, Calandra P, Fruscoloni P, Tocchini-Valentini GP. Two helices plus a linker: a small model substrate for eukaryotic RNase P. PNAS 1995; 92. 2627-2631.

Carstens RP, Wagner EJ, Garcia-Blanco MA. An intronic splicing silencer causes skipping of the IIIb exon of fibroblast growth factor receptor 2 through involvement of polypyrimidine tract binding protein. Mol Cell Biol 2000; 20:7388-7400.

Carter MS, Kuhn KM, Sarnow P . Cellular internal ribosome entry site elements and the use of cDNA microarrays in their investigation. In: Sonenberg N, Hershey JWB, and Mathews MB (eds) Translational ControL, CSHL PRESS 2000; 615-635.

Casey JL, Bergmann KF, Brown TL, Gerin JL. Structural requirements for editing in hepatitis d virus: evidence for a uridine-to-cytidine editing mechanism. Proc Natl Acad Sci USA 1992; 89: 7149-7153.

Casey JL, Hentze MW, Koeller DM, Caugham SW, Rouault TA, Klausner RD, Harford JB. Iron-responsive elements: regulatory RNA sequences that control mRNA levels and translation. Science 1998; 240: 924-928.

Castanotto D, Rossi J. Sarver N. Antisense cataltyic RNAs as therapeutic agents. Adv Pharmacol 1994; 25: 289-317.

Cavaillé J, Nicoloso M, Bachellerie J-P. Targeted ribose methylation of RNA *in vivo* directed by tailored antisense RNA guides. Nature 1996; 383: 732-735.

Cech TR. Ribozymes and their medical implications. J Am Med Assoc 1988; 260: 3030-3034.

Cech TR. Structure and mechanisms of the large catalytic RNAs: groupI and group II introns and ribonuclease P. In: Gesteland RF, Atkins JF (eds), The RNA World, CSHL Press 1993; 239-269.

Cech TR. Structural biology. The ribosome is a ribozyme. Science 2000; 289:878-879.

Cech TR and Golden BL . Building a catalytic active site using only RNA. In: Gesteland RF, Cech TR, Atkins JF (eds) The RNA World, CSHL Press 1999; 321-349.

Chadalavada DM, Knudsen SM, Nakano S, Bevilacqua PC. A role for upstream RNA structure in facilitating the catalytic fold of the genomic hepatitis delta virus ribozyme. J Mol Biol 2000; 301: 349-367.

Chambon P. Split genes. Sci Am 1981; 244: 60-71.

Chang BJ, Lau P, Chan L. Apolipoprotein B mRNA editing. In: Grosjean H, Benne R (eds). Modification and Editing of RNA 1998, 325-342..

Chang FL, Chen PJ, Tu SJ, Wang CJ, Chen DS. The large form of hepatitis d antigen is crucial for assembly of hepatitis d virus. PNAS 1991; 88: 8490-8494.

Chang KY, Tinoco I, Jr. The structure of an RNA hairpin complex of the HIV TAR hairpin loop and its complement. J Mol Biol 1997; 269: 52-66.

Chao M, Hsieh SY, Taylor J. Role of two forms of the hepatitis delta virus antigen: evidence for a mechanism of self-limiting genome replication. J Virol 1990; 64: 5066-5069.

Chapdelaine Y, Bohnen L. The wheat mitochondrial gene for subunit I of the NADH dehydrogenase complex: a *trans*-splicing model for this gene-in-pieces. Cell 1991; 65: 465-472.

Chapon C, Cech TR, Zaug AJ. Polyadenylation of telomerase RNA in budding yeast. RNA 1997; 3: 1337-1351.

Chen CY, Xu N, Shyu AB. mRNA decay mediated by two distinct AU-rich elements from c-fos and granulocyte macrophage colony-stimulating factor transcripts: Different deadenylation kinetics and uncoupling from translation. Mol Cell Biol 1995; 15: 5777-5788.

Chen CY, Del Gatto-Konczak F, Wu Z, Karin M. Stabilization of interleukin-2 mRNA by the c-Jun NH2-terminal kinase pathway. Science 1998; 280: 1945-1949.

Chen JL, Blasco MA, Greider CW Secondary structure of vertebrate telomerase RNA. Cell 2000; 100 503-514.

Chen J-H, Le S-Y, Maizel J. A procedure for RNA pseudoknot prediction. Comp Appl Biosci 1992; 8: 243-248.

Chen S-H, Habib G, Yang C-Y, Gu Z-W, Lee BR, Weng S-A, Silbermann SR, Cai S-J, Deslypere JP; Rosseneu M, Gotto Jr AM, Li W-H, Chan L. Apolipoprotein B-48 is the product of a messenger RNA with an organ-specific in-frame stop codon. Science 1987; 238: 363-366.

Chiba Y, Ishikawa M, Kijima F, Tyson RH, Kim J, Yamamoto A, Nambara E, Leustek T, Wallsgove RM, Naito S. Evidence for autoregulation of cystathione gamme-synthase mRNA stability in *Arabidopsis*. Science 1999; 286:1371-1374.

Choo Y, Sánchez-Garcia I, Klug A. *In vivo* represssion by a site-specific DNA-binding protein designed against an oncogenic sequence. Nature 1994; 372: 642-645.

Chou JH, Greenberg JT, Demple B. Posttranscriptional repression of *Escherichia coli* OmpF protein in response to redox stress: postive control of the micF antisense RNA by the soxRS locus. J Bacteriol 1993; 175: 1026-1031.

Chow LT, Gelinas RE, Broker TR, Roberts RJ. An amazing sequence arrangement at the 5' ends of adenovirus 2 messenger RNA. Cell 1977; 12: 1-8.

Chu E *et al.* Autoregulation of human thymidilate synthase messenger RNA translation by thymidilate synthase. PNAS 1991; 8: 8977-8981.

Chu S, Archer RH, Zengel JM and Lindahl L. The RNA of RNase MRP is required for normal processing of ribosomal RNA. PNAS 1994; 91: 659-663.

Clark RE. Poor cellular uptake of antisense oligodeoxynucleotides: an obstacle to their use in chronic myeloid leukaemia. Leuk Lymphoma 1995; 19: 189-195.

Clemson CM, McNeil JA, Willard H, Lawerence JB. XIST paints the inactive X chromosome at interphase: evidence for a novel RNA involved in nuclear/chromosome structure. J Cell Biol 1996; 132: 1-17.

Cohen SP, McMurray LM, Levy SB. marA locus causes decreased expression of OmpF porin in multiple-antibiotic-resistant (Mar) mutants of *Escherichia coli*. J Bacteriol 1988; 170: 5416-5422.

Coles H. Nobel panel rewards prion theory after years of heated debate. Nature 1997; 389: 529.

Colomer R, Lupu R, Bacus SS, Gelmann EP. erbB-2 antisense oligonucleotides inhibit the proliferatin of breast cancer carcinoma cells with erbB-2 oncogene amplification. Br. J. Cancer 1994; 70: 819-825.

Conn GL, Draper DE, Lattman EE, Gittis AG. Crystal structure of a conserved ribosomal protein-RNA complex. Science 1999; 284: 1171-1174.

Conrad R, Thomas J, Spieth J, Blumenthal T. Insertion of part of an intron into the 5' untranslated region of a *Caenorhabditis elegans* gene converts it into a *trans*-spliced gene. Mol Cell Biol 1991; 11: 1921-1926.

Constancia M, Dean W, Lopes S, Moore T, Kelsey G, Reik W. Deletion of a silencer element in igf2 results in loss of imprinting independent of H19. Nat Genet 2000; 26: 203-206.

Cooper JP, Nimmo ER, Allshire RC, Cech T. Regulation of telomere length and function by a myb-domain protein in fission yeast. Nature 1997; 385: 744-747.

Copeland PR; Fletcher JE, Carlson BA, Hatfield DL, Driscoll DM. A novel RNA binding protein, SPB2, is required for the translation of mammalian selenoprotein mRNAs. EMBO J 2000; 19: 306-314.

Costa M, Dème E, Jacquier A, Michel F. Multiple tertiary interactions involving domain II of group II self-splicing introns. J Mol Biol 1997; 267: 520-536.

Coughlin BC, Teixeira SM, Kirchhoff LV, Donelson JE. Amastin mRNA abundance *in Trypanosoma cruzi* is controlled by a 3'-untranslated region position-dependent cis-element and an untranslated region-binding protein. J Biol Chem 2000; 275: 12051-12060.

Covello PS, Gray MW. RNA editing in plant mitochondria. Nature 1989; 341: 662-666.

Covello PS, Gray MW. RNA sequence and the nature of the Cu_A-binding site in cytochrome c oxidase. FEBS Lett 1990; 268: 5-7.

Covello PS, Gray MW. Silent mitochondrial and active nuclear genes for subunit 2 of cytochrome c oxidase (cox2) in soybean: evidence for RNA-mediated gene transfer. EMBO J 1992; 11: 3815-3820

Cox TC, Bawden MJ, Martin A, May BK. Human erythroid 5-aminolevulinate synthase: promoter analysis and identification of an iron-responsive element in the mRNA. EMBO J 1991; 10: 1891-1902.

Crick FHC. The genetic code. Cold Spring Harbor Symp Quant Biol. 1965; 31.

Crick FHC. The central dogma of molecular biology. Nature 1970; 227: 561-563.

Crick FHC. Split genes and RNA splicing. Science 1979; 204: 264-271.

Culbertson MR. RNA surveillance. Unforeseen consequences for gene expression, inherited genetic disorders and cancer. Trends Genet 1999; 15: 74-80.

Cullen BR. RNA-sequence-mediated gene regulation in HIV-1. Infect Agents Dis 1994; 3: 68-76.

Cullen BR, Malim MH. The HIV-1 Rev protein: prototype of a novel class of eukaryotic post-transcriptional regulators. TIBS 1991; 16: 346-350.

Culver GM, Cate JH, Yusupova GZ, Yusupova MM, Noller HF. Indentification of an RNA-protein bridge spanning the ribosomal subunit interace. Science 1999; 285: 2133-2136.

Currey KM, Shapiro BA. Secondary structure computer prediction of the poliovirus 5' non-coding region is improved by a genetic algorithm . CABIOS 1997; 13: 1-12.

Czaplinski K, Ruiz-Echevarria MJ, Gonzalez CI, Peltz SW. Should we kill the messenger? The role of the surveillance complex in translation termination and mRNA turnover. Bioessays 1999; 21: 685-696.

Dabeva MD, Post-Beitenmiller MA, Warner JR. Autogenous regulation of splicing of the transcript of a yeast ribosomal protein gene. Proc Natl Acad Sci USA 1986; 83:5854-5857.

Dabiri GA, Lai F, Drakas RA, Nishikura K. Editing of the GluR-B ion channel RNA *in vitro* by recombinant double-stranded RNA adenosine deaminase. EMBO J 1996; 15: 34-45.

Dalmay T, Hamilton A, Rudd S, Angell S, Baulcombe DC. An RNA-dependent RNA polymerase gene in Arabidopsis is required for posttranscriptional gene silencing mediated by a transgene but not by a virus. Cell 2000; 101: 543-53.

Dandekar T. Yeast U3 localization and correct sequence (snR17a) and promotor activity (snR17b) identified by homology search. DNA-Sequence. 1991; 1: 217-218

Dandekar T, Argos P. The GCN4 basic region leucine zipper binds DNA as a dimer of uninterrupted a-helices: Crystal structure of the protein DNA-complex (commentary). Chemtracts 1993; 4: 62-65.

Dandekar T, Argos P.*In vivo* repression by a site-specific DNA-binding protein designed against an oncogenic sequence (commentary). Chemtracts 1995; 5: 314-316.

Dandekar T, Argos P. Reconstructing the evolutionary history of the artiodactyl ribonuclease superfamily (commentary). Chemtracts 1995; 5: 324-325.

Dandekar T, Argos P. Selection of RNA-binding peptides *in vivo* (commentary). Chemtracts 1996; 6. 107-110.

Dandekar T, Argos P. Determination of the fold of the core protein of hepatitis B virus by electron cryomicroscopy (commentary). Chemtracts 1998; 11: 161-165.

Dandekar T, Hentze MW. Finding the hairpin in the haystack: Searching for RNA motifs. Trends Genetics 1995; 11: 45-50.

Dandekar T, Koch G. DNA and mRNA sequence of the immune protective DNA ligase I gene match the rev response element of HIV. DNA Sequence - the Journal of DNA sequencing and mapping 1996; 6: 119-121.

Dandekar T, Ribes V, Tollervey D. *Schizosaccharomyces pombe* U4 small nuclear RNA closely resembles vertebrate U4 and is required for growth.Journal of Molecular Biology 1989; 208:371-379.

Dandekar T, Sharma K. Regulatory RNA. Springer Verlag Berlin Heidelberg New York 1998.

Dandekar T., Sibbald PR. Trans-splicing of pre-mRNA is predicted to occur in a wide range of organisms including vertebrates. Nucleic Acids Res 1990; 18: 4719-4726.

Dandekar T, Tollervey D. Cloning of *Schizosaccharomyces pombe* genes encoding the U1,U2,U3 and U4 snRNAs. Gene 1989; 81: 227-235.

Dandekar T, Tollervey D. 33 nucleotides including the TATA box are both necessary and sufficient for U2 transcription in *Schizosaccharomyces pombe*. Molecular Microbiology 1991; 5: 1621-1625.

Dandekar T, Tollervey D. Mutational analysis of *Schizosaccharomyces pombe* U4 snRNA by plasmid exchange. Yeast 1992; 8: 647-653.

Dandekar T, .Tollervey D. Identification and functional analysis of a novel yeast small nucleolar RNA. Nucleic Acids Res 1993; 21: 5386-5390.

Dandekar T, Stripecke R, Gray NK, Goosen B, Constable A, Johansson HE, Hentze MW. Identification of a novel iron-responsive element in murine and human erythroid delta-aminolevulinic acid synthase mRNA. EMBO J 1991; 10: 1903-1909.

Dandekar T, Beyer K, Bork P, Kenealy MR, Pantopoulos K, Hentze M, Sonntag-Buck V, Flouriot G, Gannon F, Schreiber S. Systematic genomic screening and analysis of mRNA in untranslated regions and mRNA precursors: combining experimental and computational approaches. Bioinformatics 1998; 14: 271-278.

Dandekar T, Huynen M, Regula JT, Ueberle B, Zimmermann CU, Andrade MA, Doerks T, Sanchez-Pulido L, Snel B, Suyama M, Yuan YP, Herrmann R, Bork P. Re-annotating the mycoplasma pneumoniae genome sequence: adding value, function and reading frames. Nucleic Acids Res 2000; 28: 3278-3288.

Davidson NO, Powell LM, Wallis SC, Scott J. Thyroid hormone modulates the introduction of a stop codon in rat liver apolipoprotein B messenger RNA. J Biol Chem 1988; 263: 13482-13485.

Decker CJ, Parker R. Mechanisms of mRNA degradation in eucaryotes. TIBS 1994; 19: 336.

Decker CJ, Parker R. Diversity of cytoplasmic functions for the 3' untranslated region of eukaryotic transcripts. Curr Opin Cell Biol 1995; 7: 386-392.

Deckman IC, Draper DE. S4-alpha mRNA translation regulation complex. II. Secondary structure of the RNA regulatory site in the presence and absence of S4. J Mol Biol 1987; 196: 323-332.

De Guzman RN, Wu ZR, Stalling CC, Pappalardo L, Borer PN, Summers MF. Structure of the HIV-1 nucleocapsid protein bound to the SL3 psi-RNA recognition element. Science 1998; 279: 384-388.

de Lange T. Activation of telomerase in a human tumor. PNAS 1994; 91: 2882-2885.

Delbecq P, Werner M, Feller A, Filipowski RK, Messenguy F, Pierard A. A segment of mRNA encoding the leader peptide of CPA1 gene confers repressin by arginine on a heterologuos yeast gene transcript. Mol Cell Biol 1994; 14: 2378-2390.

Delihas N. Regulation of gene expression by trans-encoded antisense RNAs. Mol Microbiol 1995; 15:411-414.

De Los Santos T, Schweizer J, Rees CA, Francke U. Small evolutionary conserved RNA, resembling C/D box small nucleolar RNA, is transcibed from PWCR1, a novel imprinted gene in the Prader-Willi deletition region, which is highly expressed in brain. Am J Hum Genet 2000; 67: 1067-1082.

DeRosier DJ. Who needs crystals anyway ? Nature 1997; 386: 26-27.

Devereux J, Haeberli P, Smithies O. A comprehensive sequence analysis package for the VAX. Nucleic Acids Res 1984; 12: 387-395.

Dichtl B, Tollervey D. Pop3 is essential for the activity of the RNAse MRP and RNAse P ribonucleoproteins *in vivo*. EMBO J 1997; 16: 417-429.

Dickson K.S, Bilger A, Ballantyne S, Wickens MP. The cleavage and polyadenylation specificity factor in *Xenopus leavis* oocytes is a cytoplasmic factor involved in regulated polyadenylation. Mol Cell Biol 1999; 19: 5707-5717.

Dieci G, Percudani R, Giuliodori S, Bottarelli L, Ottonello S. TFIIIC-independent in vitro transcription of yeast tRNA genes. J Mol Biol 2000; 299: 601-613.

Diwa A, Bricker AL, Jain C, Belasco JG. An evolutionary conserved RNA stem-loop functions as a sensors that directs feedback regulation of RNase E gene expression. Genes Dev 2000; 14: 1249-1260.

Dix DJ, Lin PN, McKenzie AR, Walden WE, Theil EC. The influence of the base-paired flanking region on structure and function of the iron regulatory element. J Mol Biol 1993; 231: 230-240.

Domeier ME, Morse DP, Knight SW, Portereiko M, Bass BL, Mango SE. A link between RNA interference and nonsense-mediated decay in *Caenorhabditis elegans*. Science 2000; 289:1928-1931.

Domier LL, McCoppin NK, D'Arcy CJ. Sequence requirements for translation initiation of Rhopalosiphum padi virus ORF2. Virology 2000; 268: 264-271.

Dominski Z, Marzluff WF . Formation of the 3'end of histone mRNA. Gene 1999; 239: 1-14.

Doolittle RF. (ed) Molecular evolution: computer analysis of protein and nucleic acid sequences. Methods in enzymology vol. 183. Academic Press, San Diego, 1990.

Doolittle WF. Phylogenetic classification and the universal tree. Science 1999; 284: 2124-2129.

Doudna JA, Grosshans C, Gooding A, Kundrot CE. Crystallization of ribozymes and small RNA motifs by a sparse matrix approach. PNAS 1993; 90: 7829-7833.

Doye V, Hurt EC. Genetic approaches to nuclear pore structure and function. Trends Genet. 1995; 11 : 193-199.

Draper DE. Translation regulation of ribosomal proteins in *E.coli*. In: Ilan J (ed) Translational regulation of gene expression. Plenum Press New York, 1-25.

Draper DE. Protein-RNA recognition. Annu Rev Biochem 1995; 64: 593-620.

Dreyfuss G, Matunis MJ, Piñol-Roma S, Burd CG. hnRNP proteins and the biogenesis of mRNA. Annu Rev Biochem 1993; 62 :289-321.

Duarte CM, Pyle AM. Stepping through an RNA structure: A novel approach to conformational analysis. J Mol Biol 1998; 284: 1465-1478.

Duncker BP, Koops MD, Walker VK, Davies PL. Low temperature persistence of type I antifreeze protein is mediated by cold-specific mRNA stability. FEBS 1995; 377:185-188.

Dzau VJ, Mann MJ, Morishita R, Kaneda Y. Fusigenic viral liposome for gene therapy in cardiovascular diseases. PNAS 1996; 93: 11421-11425.

Ebeling S, Kundig C, Hennecke H. Discovery of a rhizobial RNA that is essential for symbiotic root nodule development. J Bacteriol 1991; 173: 6373-6382.

Ecker DJ, Vickers TA, Bruice TW. Pseudo-half-knot formation with RNA. Science 1992; 257: 958-961.

Eddy SR, Durbin R. Nucl Acids Res 1994; 22: 2079-2088.

Eden S, Cedar H Action at a distance. Nature 1995; 375: 16-17.

Egebjerg J, Kakekov V, Heinemann SF. Intron sequence directs RNA editing of the glutamate receptor subunit GluR2 coding sequence. PNAS 1994; 91: 10270-10274.

Ehricht R, Ellinger T, McCaskill JS. Cooperative amplification of templates by cross-hybridization (CATCH). Eur J Biochem 1997; 243: 358-364.

Eichler DC, Craig N. Processing of eukaryotic ribosomal RNA. Progr Nucl Acid Res Mol Biol 1994; 49: 197-239.

Endo T, Nadal-Ginard B. Three types of mucscle-specific gene expression in fusion-blocked rat skeletal muscle cells: Translational control in EGTA-treated cells. Cell 1987; 49: 515-526.

Ephrussi A, Lehmann R. *Oskar* induces germ cell formation. Nature 1992; 358:387-392.

Erard M, Barker DG, Amalric F, Jeang KT, Gatignol A. An Arg/Lys-rich core peptide mimics TRBP binding to the HIV-1 TAR RNA upper-stem/loop. J Mol Biol 1998; 279: 1085-1099.

Erdélyi M, Michon A-M, Guichet A, Bogucka-Glotzer J, Ephrussi A. A requirement for *Drosophila* cytoplasmic tropomyosin in *oskar* mRNA localization. Nature 1995; 377: 524-527.

Ermolaeva MD, Khalak HG, White O, Smith HO, Salzberg SL. Prediction of transcription terminators in bacterial genomes. J Mol Biol 2000; 301: 27-33.

Estevez AM, Simpson L. Uridine insertion/deletion RNA editing in trypanosome mitochondria-a review. Gene 1999; 240: 247-260.

Etzold T, Ulyanov A, Argos P. SRS: information retrieval system for molecular biology data banks. Methods Enzymol 1996; 266: 114-128.

Eul J,Graessmann M, Graesmann A. *Trans*-splicing and alternative-tandem-cis-splicing: two ways by which mammalian cells generate a truncated SV40 T-antigen. Nucleic Acids Res 1996; 24: 1653-1661.

Fabbri S, Fruscoloni P, Bufardeci E, Di Nicola Negri E, Baldi MI, Attardi DG, Mattoccia E, Tocchini-Valentini GP. Conservation of substrate recognition mechanisms by tRNA splicing endonucleases. Science 1998; 280:284-286.

Fabre E, Boelens WC, Wimmer C. Nup145p is required for nuclear export of mRRNA and binds homopolymeric RNA *in vitro* via a novel conserved motif. Cell 1994; 78: 275-289.

Fabrizio P, Abelson J. Two domains of yeast U6 small nuclear RNA required for both steps of nuclear precursor messenger RNA splicing. Science 1990; 250: 404-409.

Farabaugh PJ, Qian Q, Stahl G. Programmed translational frame shifting, hopping, and read-through of termination codons. In: Sonenberg N, Hershey JWB, Mathews MB (eds) Translational Control. CSHL Press 2000; 741-761.

Faubladier M, Cam K, Bouche J-P. *Escherichia coli* cell division inhibitor DicF-RNA of the dicB operon. J Mol Biol 1990; 212: 461-471.

Feagin JE, Stuart JM. Developmental aspects of uridine addition within mitochondrial transcripts of *Trypanosoma brucei*. Mol Cell Biol 1988; 8: 1259-1265.

Feagin JE, Jasmer DP, Stuart K. Developmentally regulated addition of nucleotides within apocytochrome b transcripts in *Trypanosoma brucei*. Cell 1987; 49: 337-345.

Fedor MJ. Structure and function of the hairpin ribozyme. J Mol Biol 2000; 297:269-291.

Felip E, Del-Campo JM, Rubio D, Vidal MT, Colomer R, Bermejo B. Overexpression of c-erbB-2 in epithelial ovarian cancer. prognostic value and relationship with response to chemotherapy. Cancer 1995; 75: 2147-2152.

Feng J, Funk W, Wang S., Weinrich SL, Avilion AA, Chiu CP, Adams R.R, Chang E, Allsopp R, Yu S, Le S, West D, Harley CB, Andrews WH, Greider CW, Villeponteau B. The RNA component of human telomerase. Science 1995; 269: 1236-1241.

Ferguson KC, Heid PJ, Rothman JH. The SL1 *trans*-spliced leader RNA performs an essential embryological function in *Caenorhabditis elegans* that can also be supplied by SL2 RNA. Genes Dev 1996; 10:1543-1556.

Ferguson-Smith AC, Sasaki H, Cattanach BM, Surani MA. Parental-origin-specific epigenetic modification of the mouse H19 gene. Nature 1993; 362: 751-755.

Ferrandon D, Elphick L, Nüsslein-Volhard C, St Johnston D. Staufen protein associates with the 3'UTR of bicoid mRNA to form particles that move in a microtubule-dependent manner. Cell 1994; 79: 1221-1232.

Ferre-D'Amare AR, Zhou K, Doudna JA. A general module for RNA crystallization. J Mol Biol 1998a; 279: 621-631.

Ferre-D'Amare AR, Zhou K, Doudna JA. Crystal structure of a hepatitis delta virus ribozyme. Nature 1998b; 395: 567-574.

Fey J, Tomita K, Bergdoll M, Marechal-Drouard L. Evolutionary and functional aspects of C-to-U editing at position 28 of tRNA(Cys) (GCA) in plant mitochondria. RNA 2000; 6: 470-474.

Fichant GA, Burks C. Identification of potential tRNA genes in genomic DNA sequences. J Mol Biol 1991; 220:659-671.

Flach J, Bossie M, Vogl J, Corbett A, Jinks T, Williams DA, Silver PA. A yeast RNA-binding protein shuttles between the nucleus and the cytoplasm. Mol Cell Biol 1994; 14: 8399-8407.

Fontana W, Schuster P. Continuity in evolution: on the nature of transitions. Science 1998; 280: 1451-1455.

Forchhammer K, Leinfelder W, Boeck A. Identification of a novel translation factor necessary for the incorporation of selenoysteine into protein. Nature 1989; 342: 453-456.

Forlan N, Martegan E, Alberghina L. Post transcriptional regulation of MET2 gene of *S.cerevisiae*. Biochem Biophys Acta 1991; 1089: 47-53.

Fox CA, Sheets MD, Wickens MP. Poly(A) addition during maturation of frog oocytes: distinct nuclear and cytoplasmic activities and regulation by the sequence UUUUUAU. Genes Dev 1989; 3: 2151-2162.

Franch T, Petersen M, Wagner EG, Jacobsen JP, Gerdes K. Antisense RNA regulation in prokaryotes: rapid RNA/RNA interaction facilitated by a general U-turn loop structure. J Mol Biol 1999; 294: 1115-1125.

Fritz CC, Zapp ML, Green MR. A human nucleoporin like protein that specifically interacts with Hiv Rev. Nature 1995; 376: 530-533.

Gan W, Rhoads RE. Internal initiation of translation directed by the 5' untranslated region of the mRNA for eIF4G, a factor involved in the picornavirus-induced switch from cap-dependent to internal initiation. J Biol Chem 1996; 271: 623-626.

Gan W, LaCelle M, Rhoads RE. Functional characterization of the internal ribosome entry site of eIF4G mRNA. J Biol Chem 1998; 273: 5006-5012.

Ganot P, Bortolin M-L, Kiss T. Site-specific pseudouridine formation in preribosomal RNA is guided by small nucleolar RNAs. Cell 1997; 89: 799-809.

Garcia-Barrio MT, Naranda T, Vazquez-de-Aldana CR, Cuesta R, Hinnebusch AG, Tamame M. GCD10, a translational repressor of GCN4, is the RNA-binding subunit of eukaryotic initiation factor-3. Genes Dev 1995; 9: 1781-1796.

Gaspin C, Cavaille J, Erauso G, Bachellerie JP. Archeal homologs of eukaryotic methylation guide small nucleolar RNAs: lessons from the Pyrococcus genomes. J Mol Biol 2000; 297: 895-906.

Gautheret D, Cedergren RJ. Modeling the three dimensional structure of RNA using discrete nucleotide conformation sets. Mol Biol 1993; 229: 1049-1064.

Gay DA, Sisodia SS, Cleveland DW. Autoregulatory control of beta-tubulin mRNA stability is linked to translation elongation. PNAS 1989; 86: 5763-5767.

Geballe AP, Spaete RR, Mocarski ES. A *cis*-acting element within the 5'leader of a cytomegalovirus ß transcript determines kinetic class. Cell 1986; 46: 865-872.

Gebauer F, Richter JD. Cloning and characterization of a *Xenopus*poly(A) polymerase. Mol. Cell. Biol. 1995; 15: 1422-1430.

Gebauer F, Corona DFV, Preiss T, Becker PB, Hentze MW . Translational control of dosage compensation in *Drosophila* by sex-lethal: cooperative silencing via the 5' and 3'UTRs of msl-2 mRNA is independent of the poly(A) tail. EMBO J 1999; 18: 6146-6154.

Gebauer F, Merendino L, Hentze MW, Valcarcel J. Novel functions for nuclear factors in the cytoplasm: the Sex-lethal paradigm. Semin Cell Dev Biol 1997; 8: 561-566.

Gesteland RF, Atkins JF (eds). The RNA World. 1993; CSHL Press.

Gesteland RF, Cech TR, Atkins JF (eds). The RNA World. 1999; CSHL Press.

Ghisolfi-Nieto L, Joseph G, Puvion-Dutilleul F, Amalric F, Bouvet P. Nucleolin is a sequence-specific RNA-binding protein: characterization of targets on pre-ribosomal RNA. J Mol Biol 1996; 260: 34-53.

Giddings MC, Matveeva OV, Atkins JF, Gesteland RF. ODNBase-a web database for antisense oligonucleotide effectiveness studies. Bioinformatics 2000; 16: 843-844.

Gilbert W, de Souza SJ . Introns and the RNA world. In: Gesteland RF, Cech TR, Atkins JF (eds.) The RNA World. 1999; CSHL Press: 221-231.

Gilley D, Blackburn EH. The telomerase RNA pseudoknot is critical for the stable assembly of a catalytically active ribonucleoprotein. PNAS 1999; 96: 6621-6625.

Gilmartin GM, Fleming ES, Oetjen J, Graveley BR. CPSF recognition of an HIV-1 mRNA 3'-processing enhancer: multiple sequence contacts involved in poly(A) sited definition. Genes Dev 1995; 9: 72-83.

Gingras AC. Raught B, Sonenberg N. eIF4 initiation factors: effectors of mRNA recruitment to ribosomes and regulators of translation. Annu Rev Biochem 1999; 68: 913-963.

Girard JP, Lehtonen H, Caizergues-Ferrer M, Amalric F, Tollervey D, Lapeyre B. GAR1 is an essential small nucleolar RNP protein required for pre- rRNA processing in yeast. EMBO J 1992; 11:673-682.

Giuliano F, Arrigo P, Scalia F, Cardo PP, Damiani G. Potentially functional regions of nucleic acids recognized by a Kohonen's self-organizing map. Comput Appl Biosci 1993; 9: 687-693.

Glenn JS, White JM. trans-dominant inhibition of human hepatitis delta virus genome replication. J Virol 1991; 65: 2357-2361.

Gohlmann HW, Weiner J 3rd, Schon A, Herrmann R. Identification of a small RNA within the pdh gene cluster of *Mycoplasma pneumoniae* and *Mycoplasma genitalium*. J Bacteriol 2000; 182: 3281-3284.

Golden BL, Gooding AR, Podell ER, Cech TR. A preorganized active site in the crystal structure of the *Tetrahymena* ribozyme. Science 1998; 282: 259-264.

Goldschmidt-Clermont M, Choquet Y, Girard-Bascou J, Michel F, Schirmer-Rahire M, Rochaix JD. A small chloroplast RNA may be required for *trans*-splicing in *Chlamydomonas reinhardtii*. Cell 1991; 65: 135-143.

Golightly LM, Mbacham W, Daily J, Wirth DF. 3'UTR elements enhance expression of Pgs28, an ookinete protein of *Plasmodium gallinaceum*. Mol Biochem Parasitol 2000; 105: 61-70.

Gollnick P. Regulation of *Bacillus subtilis* trp operon by an RNA-binding protein. Mol Microbiol 1994; 11: 991-997.

Goodwin EB, Okkema PG, Evans TC, Kimble J. Translation al regulation of tra-2 by its 3' untranslated region controls sexual identity in *C. elegans*. Cell 1993; 75: 329-339.

Gopal V, Brieba LG, Guajardo R, McAllister WT, Sousa R. Characterization of structural features important for T7 RNAP elongation complex stability reveals competing complex conformations and a role for the non-template strand in RNA displacement. J Mol Biol 1999; 290: 411-431.

Görlich D, Mattaj IW. Protein kinesis - nucleocytoplasmic transport. Science 1996; 271: 1513-1518.

Gott JM, Emeson RB. Functions and mechanisms of RNA editing. Annu Rev Genet 2000; 34: 499-531.

Gottlien E. The 3' untranslated region of localized maternal messages contains a conserved motif involved in mRNA localization. PNAS 1992; 89: 7164-7168.

Gouet P, Diprose JM, Grimes JM, Malby R, Burroughs JN, Zientara S, Stuart DI, Mertens PP. The highly ordered double-stranded RNA genome of bluetongue virus revealed by crystallography. Cell 1999; 97: 481-490.

Graber JH, Cantor CR, Mohr SC, Smith TF. In silico detection of control signals: mRNA 3'-end processing sequences in different species. PNAS 1999; 96: 14055-14060.

Grant CM, Miller PF, Hinnebusch AG. Sequences 5' of the first upstream open reading frame in GCN4 mRNA are required for efficient translational reinitiation. Nucleic Acids Res 1995; 23: 3980-3988.

Grassi G, Marini JC. Ribozymes: structure, function, and potential therapy for dominant genetic disorders. Ann Med 1996; 28: 499-510.

Graves LE, Segal S, Goodwin EB. TRA-1 regulates the cellular distribution of the tra-2 mRNA in C. elegans. Nature 1999; 399: 802-805.

Gray MW, Covello PS. RNA editing in plant mitochondria and chloroplasts. FASEB J 1993; 7: 64-71.

Gray MW, Hanic-Joyce PJ, Covello PS. Transcription, processing and editing in plant mitochondria. Annu Rev Plant Phys Mol Biol 1992; 43: 145-175.

Gray NK, Hentze MW. Iron regulatory protein prevents binding of the 43S translation pre-initiation complex to ferritin and eALAS mRNAs. EMBO J 1994; 13: 3882-3891.

Gray NK, Costas P, Dandekar T, Ackrell BAC, Hentze MW. Translational regulation of mammalian and *Drosophila* citric acid cycle enzymes via iron-responsive elements. PNAS 1996; 93: 4925-4930.

Green MR. Biochemical mechanisms of constitutive and regulated pre-mRNA splicing. Annu Rev Cell Biol 1991; 7: 559-599.

Green R, Noller HF. Ribosomes and translation. Annu Rev Biochem 1997; 66: 679-716.

Green PL, Yip MT, Xie Y, Chen IS. Phosphorylation regulates RNA binding by the human T-cell leukemia virus Rex protein. J Virol 1992; 4325-4330.

Greider CW, Blackburn EH. Identification of a specific telomer terminal transferase activity in *Tetrahymena* extracts. Cell 1985; 43: 405-413.

Greider CW, Blackburn EH. The telomer terminal trasnferase of tetranhymena is a ribonucleoprotein enzyme with two kinds of primer specificity. Cell 1987; 51: 887-898.

Greider CW, Blackburn EH. A telomeric sequence in the RNA of *Tetrahymena* telomerase required for telomere repeat syntheis. Nature 1989; 337: 331-337.

Grens A, Scheffler IE. The 5'- and 3'-untranslated regions of ornithine-decarboxylase mRNA affect the translational efficiency. J Biol Chem 1990; 265: 11810-11816.

Grillo G, Attimonelli M, Liunni S, Graziano P. CLEANUP: a fast computer program for removing redundancies from nucleotide sequence databases. CABIOS 1996; 12: 1-8.

Grishok FJ, Tabara H, Mello CC. Genetic requirements for inheritance of RNAi in C. elegans. Science 2000; 287: 2494-2497.

Grundy FJ, Henkin TM. The rps gene encoding ribosomal protein S4 is autogenously regulated in *Bacillus subtilis*. J Bacteriol 1991; 173: 4595-4602.

Gu H, Das Gupta J, Schoenberg DR. The poly(A) limiting element is a conserved cis-acting sequence that regulates poly(A) tail length on nuclear pre-mRNA. PNAS 1999; 96: 8943-8948.

Gualberto JM, lamattina L, Bonnard G, Weil JH, Grienenberger JM. RNA editing in wheat mitochondria results in the conservation of protein sequences. Nature 1989; 341: 660-662.

Guan H, Carpenter CD, Simon AE. Requirement of a 5'-proximal linear sequence on minus strand for plus-strand synthesis of a satellite RNA associated with turnip crinkle virus. Virology 2000; 268: 355-362.

Gultyaev AP, van Batenbergh FH, Pleij CW. The computer simulation of RNA folding pathways using a genetic algorithm. J Mol Biol 1995; 250: 37-51.

Gundelfinger ED, Krause E, Melli M, Dobberstein B. The organization of the 7SL RNA in the signal recognition particle. Nucl Acids Res 1983; 11: 7363-7374.

Guo H, Karberg M, Long M, Jones JP 3rd, Sullenger B, Lambowitz AM. Group II introns designed to insert into therapeutically relevant DNA target sites in human cells. Science 2000; 289: 452-457.

Gutell RR, Power A, Hertz GZ, Putz EJ, Stormo GD. Three dimensioanl constraints on the higher order structure of RNA: continued development and application of computational sequence analysis methods. Nucleic Acids Res 1992; 20: 5785-5795.

Gutell RR, Cannone JJ, Konings D, Gautheret D. Predicting U-turns in ribosomal RNA with comparative sequence analysis. J Mol Biol 2000; 300: 791-803.

Guthrie C. Messenger RNA splicing in yeast: clues to why the spliceosome is a ribonucleoprotein. Science 1991; 253: 157-163.

Ha I, Wightman B, Ruvkun G. A bulged lin-4/lin-14 RNA duplex is sufficient for *Caenorhabditis elegans* lin-14 temporal gradient formation. Genes Dev 1996; 10: 3041-3050.

Hajduk SL, Sabatini RS. Mitochondrial mRNA editing in kinetoplastid protozoa. In: Grosjean H, Benne R (eds) Modification and editing of RNA 1998, 377-393.

Hajduk SL, Harris ME, Pollard VW. RNA editing in kinetoplastid mitochondria. FASEB J 1993;7:54-63.

Hake LE, Richter JD. CPEB is a specificity factor that mediates cytoplasmic polyadenylation during Xenopus oocyte maturation. Cell 1994; 79: 617-627.

Hake LE, Mendez R, Richter JD. Specificity of RNA binding by CPEB: requirement for RNA recognition motifs and a novel zink-finger. Mol Cell Biol 1998; 18: 685-693.

Hall BD, Spiegelman S. Sequence complementarity of T2-DNA and T2-specific RNA. PNAS 1964; 47: 137-146.

Hall SL, Padgett RA. Conserved sequences in a class of rare eukaryotic nuclear introns with non-consensus splice sites. J Mol Biol 1994; 239: 357-365.

Hall SL, Padgett RA. Requirement of U12 snRNA for *in vivo* splicing of a minor class of eukaryotic nuclear pre-mRNA introns. Science 1996; 271: 1716-1718.

Hamilton AJ, Baulcombe DC. A species of small antisense RNA in posttranscriptional gene silencing in plants. Science 1999; 286: 950-952.

Hamilton SE, Simmons CG, Kathiriya IS, Corey DR. Cellular delivery of peptide nucleic acids and inhibition of human telomerase. Chem Biol 1999; 6: 343-351.

Hammond SM, Bernstein E, Beach D, Hannon GJ. An RNA-directed nuclease mediates post-transcriptional gene silencing in Drosophila cells. Nature 2000; 404: 293-296.

Han K, Kim HJ. Prediction of common folding structures of homologuos RNAs. Nucleic Acids Res 1993; 21: 1251-1257.

Handa N, Nureki O, Kurimoto K, Kim I, Sakamoto H, Shimura Y, Muto Y, Yokoyama S. Structural basis for recognition of the tra mRNA precursor by the Sex-lethal protein. Nature 1999;398:579-585.

Hao Y, Crenshaw T, Moulton T, Newcomb E, Tycko B. Tumour-suppressor activity of H19 RNA. Nature 1993; 365: 764-767.

Harada K, Martin SS, Frankel AD. Selection of RNA-binding peptides *in vivo*. Nature 1996; 380: 175-179.

Hardin PE, Hall JC, Rosbash M. Feedback of the *Drosophila* period gene product on circadian cycling of its mRNA levels. Nature 1990; 343: 536-540.

Harley CB, Villeponteau B. Telomeres and telomerases in aging and cancer. Curr Opin Genet Dev 1995; 5: 249-255.

Harley CB, Kim NW. Telomerase and cancer. Important Adv Oncol 1996; 1: 57-67.

Harris ME, Hajduk SL. Kinetoplastid editing: in vitro formation of cytochrome b guide RNA-mRNA chimeras from synthetic substrate RNAs. Cell 1992; 68: 1091-1099.

Hashimoto C, Steitz JA. U4 and U6 RNAs coexist in a single small nuclear ribonucleoprotein particle. Nucleic Acid Res 1984; 12: 3283-3293

Hemmings-Mieszczak M, Steger G, Hohn T. Alternative structrues of the cauliflower mosaic virus 35S RNA leader: implication for viral expression and replication. J Mol Biol 1997; 267: 1075-1088.

Hendrich BD, Willard HF Epigenetic regulation of gene expression: the effect of altered chromatin structure from yeast to mammals. Hum Mol Genet 1995; 4: 1765-1777.

Henry Y, Wood H, Morrisey JP, Petfalski E, Kearsey S, Tollervey D. The 5' end of yeast 5.8S rRNA is generated by exonucleases from an upstream cleavage site. EMBO J 1994; 13: 2452-2463.

Hentze MW, Kühn LC. Molecular control of vertebrate iron metabolism: mRNA-based regulatory circuits operated by iron, nitric oxide, and oxidative stress. PNAS 1996; 93: 8175-8182.

Hentze MW, Kulozik AE. A perfect message: RNA surveillance and nonsense-mediated decay. Cell 1999; 96: 307-310.

Hentze MW, Wright Caughman S, Rouault TA, Barriocanal JG, Dancis A, Harford JB, Klausner RD. Identification of the iron-responsive element for the translational regulation of human ferritin mRNA. Science 1987; 238: 1570-1573.

Herb A, Higuchi M, Sprengel R, Seeburg PH. Q/R site editing in kainate receptor GluR5 and GluR6 pre-mRNAs requires distant intronic sequences. PNAS 93: 1875-1880.

Hermann T, Westhof E. Aminoglycoside binding to the hammerhead ribozyme: a general model for the interaction of cationic antibiotics with RNA. J Mol Biol 1998; 276: 903-912.

Hertel KJ, Nerschlag D, Uhlenbeck OC. Specificity of hammerhead ribozyme cleavage. EMBO J 1996; 15: 3751-3757.

Herzing LBK, Romer JT, Horn JM, Ashworth A. Xist has properties of the X-chromosome inactivation center. Nature 1997; 386: 272-279.

Hickerson RP, Watkins-Sims CD, Burrows CJ, Atkins JF, Gesteland RF, Felden B. A nickel complex cleaves uridine in folded RNA structures: application to *E. coli* tmRNA and related engineered molecules. J Mol Biol 1998; 279: 577-87.

Higgins DG, Bleasby AJ, Fuchs R. Comp Appl Biosci 1992; 8: 189-191.

Higgins DG, Thompson JD, Gibson TJ. Using CLUSTAL for multiple sequence alignments. Methods Enzymol. 1996; 266: 383-402.

Higgins SJ, Hames BD (eds). RNA processing - a practical approach. Oxford University Press, 1994.

Higuchi M, Single FN, Köhler M, Sommer B, Sprengel R, Seeburg PH. RNA editing of AMPA receptor subunit GluR-B: a base-paired intron-exon structure determines position and efficiency. Cell 1993; 75: 1361-1370.

Hill KE, Lloyd RS, Burk RF. Conserved nucleotide sequences in the open reading frame and 3' untranslated region of selenoprotein P mRNA. PNAS 1993; 90: 537-541.

Himeno H, Yoshida S, Soma A, Nishikawa K. Only one nucleotide insertion to the long variable arm confers an efficient serine acceptor activity upon *Saccharomyces cerevisiae* tRNALeu *in vitro*. J Mol Biol 1997; 268: 704-711.

Hinkley CS, Blasco MA, Funk WD, Feng J, Villeponteau B, Greider CW, Herr W. The mouse telomerase RNA 5'end lies just upstream of the telomerase template sequence. Nucleic Acids Res 1998; 26:532-536.

Hinnebusch AG . Translational control of GCN4: gene-specific regulation by phosphorylation of eIF2. In: Hershey J, Mathews M, Sonenberg N (eds) Translational Control. CSHL Press 1996; 199-244.

Hinnebusch AG. Translational reguation of yeast GCN4. A window on factors that control initiator tRNA binding to the ribosome. J Biol Chem 1997; 272: 21661-21664.

Hirsch HH, Nair AP, Backenstoss V, Moroni C. Interleukin-3 mRNA stabilization by a trans-acting mechanosm in autocrine tumors lacking interleukin-3 gene rearrangements. J Biol Chem 1995; 270: 20629-20635.

Ho Y, Waring RB. The maturase encoded by a group I intron from Aspergillus nidulans stabilizes RNA tertiary structure and promotes rapid splicing. J Mol Biol 1999; 292: 987-1001.

Hobbs FW. Palladium-catalyzed synthesis of alkynylamino nucleosides. A universal linker for nucleic acids. J Org Chem 1989; 54: 3420-3422.

Hoch B, Maier RM, Appel K, Igloi GL, Kössel H. Editing of a chloroplast mRNA by creation of an initation codon. Nature 1991; 353: 178-180.

Hockenbery DM. Bcl-2, a novel regulator of cell death. Bioessays 1995; 17: 631-638.

Hodges D, Bernstein SI. Genetic and biochemical analysis of alternative RNA splicing. Academic Press, San Diego, 1994.

Hodges P, Scott J. Apolipoprotein B mRNA editing: a new tier for the control of gene expression. Trends Biochem Sci 1992; 17: 77-81.

Holcik M, Liebhaber SA. Four highly stable eukaryotic mRNAs assemble 3'UTR-protein complexes sharing *cis-* and *trans* components. PNAS 1997; 94: 2410-2414.

Holcik M, Korneluk RG. Fuctional characterisation of the X-linked inhibitor of apoptosis (XIAP) internal ribosome entry site element: role of La autoantigen in XIAP translation. Mol Cell Biol 2000; 20: 4648-4657.

Holcik M, Sonenberg N, Korneluk RG. Internal ribosome initiation of translation and the control of cell death. Trends in Genetics 2000; 16: 469-473.

Holley RW, Apgar J, Everett GA. Structure of a ribonucleic acid. Science 1965; 147: 1462-1465.

Honda M, Brown EA, Lemon SM. Stability of a stem-loop involving the initiator AUG controls the efficiency of internal initiation of translation on hepatitis C virus RNA. RNA 1996; 2: 955-968.

Hornstein E, Harel H, Levy G, Meyuhas O. Overexpression of poly(A)-binding protein down-regulates the translation or the abundance of ist own mRNA. FEBS L 1999; 457: 209-213.

Hotz HR, Lorentz P, Fischer R, Krieger S, Clayton C. Role of 3'UTRs in the regulation of hexose transporter mRNA in *T.brucei*. Mol Biochem Parasitol 1995; 75: 1-14.

Hoyne PR, Edwards LM, Viari A, Maher LJ 3rd. Searching genomes for sequences with the potential to form intrastrand triple helices. J Mol Biol 2000; 302: 797-809.

Huang H, Liao J, Cohen SN. Poly(A)- and poly(U)-specific RNA 3' tail shortening by *E. coli* ribonuclease E. Nature 1998; 391: 99-102.

Huang J, Villemain J, Padilla R, Sousa R. Mechanisms by which T7 lysozyme specifically regulates T7 RNA polymerase during different phases of transcription. J Mol Biol 1999; 293: 457-475.

Hughes-JM. Functional base-pairing interaction between highly conserved elements of U3 small nucleolar RNA and the small ribosomal subunit RNA. J Mol Biol 1996; 259: 645-654.

Hughes JMX, Ares MJ. Depletion of U3 small nucleolar RNA inhibits cleavage in the 5' external transcribed spacer of yeast pre-ribosomal RNA and impairs formation of 18S ribosomal RNA. EMBO J 1991; 10: 4231-4239.

Huttenhofer A, Boeck A. RNA structures involved in selenoprotein synthesis. In: Simon R, Grunberg-Manago, M. (eds) RNA structure and function. CSHL Press 1998, 603-639.

Huynen M, Gutell R, Konings D. Assessing the reliability of RNA folding using statistical mechanics. J Mol Biol 1997; 267:1104-1112.

Ikawa Y, Shiraishi H, Inoue T. Minimal catalytic domain of a group I self-splicing intron RNA. Nat Struct Biol 2000; 7: 1032-1035.

International Human Genome Sequencing Consortium. Initial sequencing and analysis of the human genome. Nature 2001; 409: 860-922.

Ito K, Uno M, Nakamura Y. A tripeptide 'anticodon' deciphers stop codons in messenger RNA. Nature 2000; 403: 680-684.

Iverson LE, Mottes JR, Yeager SA, Germeraad SE. Tissue-specific alternative splicing of Shaker potassim channel transcripts result from distinct modes of regulating 3' splice choice. J Neurobiol 1997; 32: 457-468.

Jackson RJ. Initiation without an end. Nature 1991; 353: 14-15.

Jackson S, Wickens M. Translational controls impiging on the 5'-untranslated region and initiation factor proteins. Curr Opin Gen Dev 1997; 7: 233-241.

Jacob F, Monod J. Genetic regulatory mechanisms in the synthesis of proteins. J Mol Biol 1961; 3: 318-356.

Jacobson A. Poly(A) metabolism and translation: the closed-loop model. In: Hershey J, Mathews M, Sonenberg N (eds) Translational Control. CSHL Press 1996; 451-480.

Jacobson A, Peltz SW . Destabilization of nonsense-containing transcripts in *S. cerevisiae*. In: Sonenberg N, Hershey JWB, and Mathews MB (eds) Translational Control. CSHL Press 2000; 827-847.

Jacquier A. Self-splicing group II and nuclear pre-mRNA introns: how similar are they? Trends Biochem Sci 1990; 15: 351-354.

Jacquier A, Jacquesson-Breuleux N. Splice site selection and the role of the lariat in a group II intron. J Mol Biol 1991; 219: 415-428.

Jan E, Yoon JW, Walterhouse D, Iannoaccone P, Goodwin B. Conservation of the C. elegans tra-2 3'UTR translational control. EMBO J. 1997; 16: 6301-6313.

Jansen R, Tollervey D, Hurt EC. A U3 snoRNP protein with homology to splicing factor PRP4 and Gb domains is required for ribosomal RNA processing. EMBO J 1993; 12: 2549-2558.

Jeanmougin F, Thompson JD, Gouy M, Higgins DG, Gibson TJ. Multiple sequence alignment with Clustal X. Trends Biochem Sci. 1998; 23:403-405.

Jentsch S. When proteins receive deadly messages at birth. Science 1996; 271: 955-956.

Ji Y, Marra A, Rosenberg M, Woodnutt G. Regulated antisense RNA eliminates alpha-toxin virulence in *Staphylococcus aureus* infection. J Bacteriol 1999; 181: 6585-90.

Jin L, Loyd RV. In situ hybridization: methods and applications. J Clin Lab Anal 1997; 11: 2-9.

Johansson HE, Sproat BS, Melefors O. Reverse transcription using nuclease-resistant primers. Nucleic Acids Res 1993; 21: 2275-2276.

Johnston CM, Nesterova TB, Formstone EJ, Newall AE, Duthie SM, Sheardown SA, Brockdorff N. Developmentally regulated Xist promoter switch mediates initiation of X inactivation. Cell 1998; 94: 809-817.

Jones EM, Gray-Keller M, Art JJ, Fettiplace R. The functional role of alternative splicing of Ca^{2+}-activated K^+ channels in auditory hair cells. Ann NY Acad Sci 1999; 868: 379-385.

Jones JT, Lee SW, Sullenger BA. Tagging ribozymes reaction sites to follow *trans*-splicing in mammalian cells. Nature Med 1996; 2: 643-648.

Jung J-E, Karoor V, Sandbaken MG, Lee BJ, Ohama T, Gesteland RF, Atkins JF, Mullenbach GT, Hill KE, Wahba AJ, Hatfield DL. Utilization of selenocysteyl-tRNA[(Ser)Sec] and seryl-tRNA[(Ser)Sec] in protein synthesis. J Biol Chem 1994; 269: 29739-29745.

Kambach C, Walke S, Young R, Avis JM, de la Fortelle E, Raker VA, Luhrmann R, Li J, Nagai K. Crystal structures of two Sm protein complexes and their implications for the assembly of the spliceosomal snRNPs. Cell 1999; 96: 375-387.

Kan JL, Green MR. Pre-mRNA splicing of IgM exons M1 and M2 is directed by a juxtaposed splicing enhancer and inhibitor. Genes Dev 1999; 13:462-471.

Kandel-Lewis S, Seraphin B. Role of U6 snRNA in 5' splice site selection. Science 1993; 262: 2035-2039.

Kang CH, Chan R, Berger I, Lockshin C, Green L, Gold L, Rich A. Crystal structure of the T4 regA translational regulator protein at 1.9 Å resolution. Science 1995; 268: 1170-1173.

Karpen GH, Schaefer JE, Laird CD. A *Drosophila* rRNA gene located in euchromatin is active in transcription and nucleolus formation. Genes Dev 1988; 2: 1745-1763.

Katsu Y, Yamashita M, Nagahama Y. Translational regulation of cyclin B mRNA by 17alpha,20beta-dihydroxy-4-pregnen-3-one (maturation-inducing hormone) during oocyte maturation I a teleost fish, the goldfish (*Carassius auratua*). Moll Cell Endocrinol 1999; 158: 79-85.

Kaufman RJ. Double-stranded RNA-activated protein kinase PKR. In: Sonenberg N, Hershey J, Mathews M (eds) Translational Control. CSHL Press 2000: 503-527

Keck JL, Roche DD, Lynch AS, Berger JM. Structure of the RNA polymerase domain of E. coli primase. Science 2000; 287: 2482-2486.

Keenan RJ, Freymann DM, Walter P, Stroud RM. Crystal structure of the signal-sequence binding subunit of the signal recognition particle. Cell 1998; 94:181-191.

Keller W. No end yet to messenger RNA 3' processing. Cell 1995; 81: 829-832.

Kelley RL, Kuroda MI.Equality for X chromosomes. Science 1995; 270: 1607-1610.

Kenealy MR, Flouriot G, Pope C, Gannon F. The 3' untranslated region of the human estrogen recpetor gene post-transcriptionally reduces mRNA levels. Biochem Soc Trans 1996; 24: 107.

Kenealy MR, Flouriot G, Sonntag-Buck V, Dandekar T, Brand H, Gannon F. The 3'-untranslated region of the human estrogen receptor alpha gene mediates rapid messenger ribonucleic acid turnover. Endocrinology 2000; 141: 2805-2813.

Ketting RF, Plasterk RH. A genetic link between co-suppression and RNA interference in *C. elegance*. Nature 2000; 404: 296-298.

Kieft JS and Tinoco I, Jr. Solution structure of a metal-binding site in the major groove of RNA complexed with cobalt (III) hexammine. Structure 1997; 5. 713-721.

Kiledjian M, Dreyfuss G. Primary structure and binding activity of the hnRNP U protein: binding RNA through RGG box. EMBO J 1992; 11: 2655-2664.

Kiledjian M, Wang X, Liebhaber SA. Identification of two KH domian proteins in the alpha-globin mRNP stability complex. EMBO J 1995; 14: 43570-4364.

Kim J, Cole JR, Pramanik S. Alignment of possible secondary structures in multiple RNA sequences using simulated annealing. CABIOS 1996; 12: 259-267.

Kim NA, Piatyszek MA, Prowse KR, Harley CB, West MD, Ho PLC, Coviello GM, Wright WE, Weinrich SL, Shay JW. Specific association of human telomerase activity with immortal cells and cancer. Science 1994; 266: 2011-2015.

Kim U, Nishikura K. Double-stranded RNA adenosine deaminase as a potential mammalian RNA editing factor. Semin Cell Biol 1993; 4: 285-293.

Kim-Ha J, Smith JL, Macdonald PM. *oskar* mRNA is localized to the posterior pole of the *Drosophila* oocyte. Cell 1991; 66: 23-35.

Kim-Ha J, Webster PJ, Smith JL, Macdonald PM. Multiple RNA regulatory elements mediate distinct steps in the localization of *oskar* mRNA. Development 1993; 119: 169-178.

Kim-Ha J, Kerr K, Macdonald PM. Translational regulation of *oskar* mRNA by bruno, an ovarian RNA-binding protein, is essential. Cell 1995; 81: 403-412.

Kiss T, Filipowicz W. Exonucleolytic processing of small nucleolar RNAs from pre-mRNA introns. Genes Dev 1995; 9: 1411-1424.

Kiss-László Z, Henry Y, Bachellereie J-P, Caizergues-Ferrer M, Kiss T. Site-specific ribose methylation of preribosomal RNA: A novel function for small nucleolar RNAs. Cell 1996; 85: 1077-1088.

Klasens BI, Thiesen M, Virtanen A, Berkhout B. The ability of HIV-1 AAUAAA signal to bind polyadenylation factors is controlled by local RNA structure. Nucleic Acid Res 1999; 27:446-454.

Klausner RD, Rouault TA, Harford JB. Regulating the fate of mRNA: The role of cellular iron metabolism. Cell 1993; 72: 19-28.

Klootwijk J, Planta RJ. Isolation and characterization of yeast ribosomal RNA precursors and preribosomes. Methods Enzymol 1989; 180: 96-109.

Knight SW, Docherty K. The identification of protein RNA interactions within the 5' UTR of human preproinsulin mRNA. Bioch Soc Trans 1991; 19: 120.

Knoop V, Schuster W, Wissinger B, Brennicke A. *Trans* splicing integrates an exon of 22 nucleotides into the nad5 mRNA in higher plant mitochondria. EMBO J 1991; 10: 3483-3493

Koch G, Dandekar T. RNA-catalysed nucleotide synthesis. Condensation and commentary. Chemtracts -Biochem Mol Biol 1999; 12: 938-943.

Kochetov AV, Ponomarenko MP, Frolov AS, Kisselev LL, Kolchanov NA. Prediction of eukaryotic mRNA translational properties. Bioinformatics 1999; 15: 704-712.

Kohler SA, Menotti E, Kühn LC. Molecular clonino of mouse gycolate oxidase. High evolutionary conservation and presence of iron-responsive element-like sequence in the mRNA. J Biol Chem 1999; 274: 2401-2407.

Kolakofsky D, Hausmann S . Cotranscriptional paramyxovirus mRNA editing: a contradiction in terms? In: Grosjean H, Benne R (eds) Modification and editing of RNA 1998, 13-420.

Kolk MH, van der Graaf M, Wijmenga SS, Pleij CW, Heus HA, Hilbers CW. NMR structure of a classical pseudoknot: interplay of single- and double-stranded RNA. Science 1998; 280: 434-438.

Komatsu Y, Yamashita S, Kazama N, Nobuoka K, Ohtsuka E. Construction of new ribozymes requiring short regulator oligonucleotides as a cofactor. J Mol Biol 2000; 299: 1231-1243.

Komine Y, Kitabatake M, Yokogawa T, Nishikawa K, Inokuchi H. A t-RNA-like structure is present in 10Sa RNA, a small stable RNA from *Escherichia coli*. PNAS 1994; 91: 9223-9227.

Konarska MM, Grabowski PJ, Padgett RA, Sharo PA. Characterization of the branch site in lariat RNAs produced by splicing of mRNA precursors. Nature 1985; 313: 552-557.

Konings DAM, Nash MA, Maizel JV, Arlinghaus RB. Novel GACG-hairpin motif in the 5'untranslated region of type (C) retroviruses related to murine leukemia virus. J Virol 1992; 66:632-640.

Koslowsky DJ, Bhat GJ, Perollaz AL, Feagin JE, Stuart K. The MURF3 gene of *T. brucei* contains multiple domains of extensive editing and is homologous to a subunit of NADH dehydrogenase. Cell 1990; 62: 901-911.

Kozak M. The scanning model for translation: an update. J Cell Biol 1989;108:229-241.

Kozak M. Interpreting cDNA sequences: some insights from studies on translation. Mamm-Genome 1996; 7: 563-574.

Krause M, Hirsch D. A *trans*-spliced leader sequence on actin mRNA in *C. elegans*. Cell 1987; 49: 753-761.

Kreivi J-P, Lamond AI. RNA splicing: Unexpected spliceosome diversity. Curr Biol 1996; 6: 802-805.

Kruys V, Wathelet M, Poupart, Contreras R, Fiers W, Content J, Huez G. The 3' untranslated region of the human interferon-ß mRNA has an inhibitory effect on translation. Proc Natl Acad Sci USA 1987; 84:6030-6034.

Kubota S, Kondo S, Eguchi T, Hattori T, Nakanishi T, Pomerantz RJ, Takigawa M. Identification of an RNA element that confers post-transcriptional repression of connective tissue growth factor/hypertrophic chondrocyte specific 24 (ctfg/hcs24) gene: similarities to retroviral RNA-protein interactions. Oncogene 2000; 19:4773-4786.

Kuchino Y, Muramatsu T. Nonsense suppression in mammalian cells. Biochimie 1996; 78: 1007-1015.

Kudla J, Igloi GL, Metzlaff M, Hagemann R, Kössel H. RNA editing in tobacco chloroplasts leads to the formation of a translatable psbL mRNA by a C to U substitution within the initiation codon. EMBO J 1992; 11: 1099-1103.

Kudo M, Kitamura-Abe S, Shimbo M, Iida Y. Analysis of 5' splice site sequences in mammalian RNA precursors by a subclass method. Comp Appl Biosci 1992; 8: 367-376.

Kufel J, Kirsebom LA. Residues in Escherichia coli RNAse P RNA important for cleavage site selection and divalent metal ion binding. J Mol Biol 1996; 263: 685-698.

Kundu M, Ansari SA, Chepenik LG, Pomerantz RJ, Khalili K, Rappaport J, Amini S. HIV-1 regulatory protein tat induces RNA binding proteins in central nervous system cells that associate with the viral *trans*-acting response regulatory motif. J Hum Virol 1999; 2:72-80.

Kuo MY, Chao M, Taylor J. Initiation of replication of the human hepatitis delta virus genome from cloned DNA: role of delta antigen. J Virol 1989; 63: 1945-1950.

Kuroda MI, Palmer MJ, Lucchesi J C. X-chromosome dosage compensation in *Drosophila*. Semin Dev Biol 1993; 4: 107-116.

Kuwabara PE, Okkema PG, Kimble J. tra-2 encodes a membrane protein and may mediate cell communication in the *Caenorhabditis elegans* sex determination pathway. Mol Cell Biol 1992; 3: 461-473.

Kwon YK, Hecht NB. Binding of a phosphoprotein to the 3'untranslated region of the mouse protamine 2 mRNA temporally represses its translation. MCB 1993; 13:6547-6557.

Kyprides NC, Ouzounis CA. Mechanisms of specificity in mRNA degradation:Autoregulation and cognate interactions. J Theor Biol 1993; 163: 373-392.

Laforest MJ, Roewer I, Lang BF. Mitochondrial tRNAs in the lower fungus *Spizellomyces punctatus*: tRNA editing and UAG 'stop' codons recognized as leucine. Nucleic Acids Res 1997; 25: 626-632.

Lai EC, Posakony JW. Regulation of *Drosophila neurogenesis* by RNA:RNA duplexes? Cell 1998; 93:1103-1104.

Laing LG, Hall KB. A model of the iron-responsive element RNA hairpin loop structure determined from NMR and thermodynamic data. Biochemistry 1996; 35:13586-13596.

Laird PW. Transsplicing in trypanosomes - archaism or adaptation? Trends Genet 1989; 5:204-209.

Lambowitz AM, Caprara MG, Zimmerly S, Perlman PS. Group I and group II ribozymes as RNPs: clues to the past and guides to the future. In: Gesteland RF, Cech TR, and Atkins JF (eds) The RNA world, 2nd edn. 1999; 451-485.

Lamond A (ed.) pre-mRNA processing. Molecular Biology Intelligence Unit, RG Landes, Austin, Texas, 1995.

Lamond AI, Earnshaw WC. Structure and function in the nucleus. Science 1998; 280:547-553.

Lamond AI, Konarska MM, Grabowski PJ, Sharp P. Spliceosome assembly involves the binding and release of U4 small nuclear ribonucleoprotein. Proc Natl Acad Sci 1988; 85: 411-415.

Landers JE, Cassel SL, George DL. Translational enhancement of mdm2 oncogene expression in human tumor cells containing a stabilized wild-type p53 protein. Cancer Res 1997; 57: 3562-3568.

Lazowska J, Jacq C, Slonimski PP. Sequence of introns and flanking exons in wild-type and box3 mutants of cytochrome b reveals an interlaced splicing protein coded by an intron. Cell 1980: 22, 333-348.

Lee CZ, Lin JH, CHao M, McKnight K, Lai MM. RNA binding activity of hepatitis delta antigen involves two arginine-rich motifs and is required for hepatitis delta virus RNA replication. J Virol 1993; 67:2221-2227.

Lee JT. Disruption of imprinted X inactivation by parent-of-origin effects at TSIX.Cell 2000; 103:17-27.

Lee JT, Jaenisch R. Long-range *cis* effects of ectopic X-inactivation centres on a mouse autosome. Nature 1997; 386: 275-279.

Lee JT, Strauss WM, Dausman JA, Jaenisch R. A 450-kb transgene displays properties of the mammalian X-inactivation center. Cell 1996; 86: 83-94.

Lee RC, Feinbaum RL, Ambros V. The *C.elegans* heterochronic gene lin-4 encodes small RNAs with antisense complementarity to lin-14. Cell 1993; 75:843-854.

Legault P, Li J, Mogridge J, Kay LE, Greenblatt J. NMR structure of the bacteriophage lambda N peptide/boxB RNA complex: recognition of a GNRA fold by an arginine-rich motif. Cell 1998; 93:289-299.

Leontis NB, Westhof E. A common motif organizes the structure of multi-helix loops in 16 S and 23 S ribosomal RNAs. J Mol Biol 1998; 283:571-583.

Lerner MR, Boyle JA, Mount SM, Wolin SL, Steitz JA. Are snRNPs involved in splicing? Nature 1980; 283: 220-224.

Lesoon A, Mehta A, Singh R, Chisolm GM, Driscoll DM. An RNA-binding protein recognizes a mammalian selenocysteine insertion sequence element required for cotranslational incorporation of selenocysteine. Mol Cell Biol 1997; 17:1977-1985.

Lewin AS, Hauswirth WW. Ribozyme gene therapy: application for molecular medicine. Trends Mol Med 2001; 7: 221-228.

Lewis HA, Musunuru K, Jensen KB, Edo C, Chen H, Darnell RB, Burley SK. Sequence specific RNA binding by a nova KH domain: implications for paraneoplastic disease and the fragile X syndrome. Cell 2000; 100:323-332.

Lewis JD, Tollervey D. Like attracts like: getting RNA processing together in the nucleus. Science 2000; 288:1385-1389.

Li E, Beard C, Jaenisch R. Role for DNA methylation in genomic imprinting. Nature 1993; 366: 362-365.

Li H, Abelson J. Crystal structure of a dimeric archaeal splicing endonuclease. J Mol Biol 2000; 302:639-648.

Li HV, ZagorskiJ, Fournier MJ. Depletion of U14 small nuclear RNA (snR128) disrupts production of 18S rRNA in *Saccharomyces cerevisiae*. Mol Cell Biol 1990; 10: 1145-1152.

Li J, Petryshyn RA. Activation of the double-stranded RNA-dependent eIF-2 alpha kinase by cellular RNA from 3T3-F442A cells. Eur J Biochem 1991; 195: 41-48.

Li S, Wilkinson MF. Nonsense surveillance in lymphocytes? Immunity 1998; 8:135-141.

Li Z, Brow DA. A spontaneous duplication in U6 spliceosomal RNA uncouples the early and late functions of the ACAGA element in vivo. RNA 1996;2:879-894.

Liang H, Jost JP. An estrogen-dependent polysomal protein binds to the 5'UTR of the chicken vitellogenin mRNA. Nucleic Acids Res. 1991; 19:2289-2994.

Lim J, Thomas T, Cavicchioli R. Low temperature regulated DEAD-box RNA helicase from the Antarctic archaeon, *Methanococcoides burtonii*. J Mol Biol 2000; 297:553-67.

Limbach PA, Crain PF, McCloskey JA. Summary: the modified nucleosides of RNA. Nucleic Acids Res 1994; 22: 2183-2196.

Lin FT, MacDougald OA, Diehl AM and Lane MD. A 30-kDa alternative translation product of the CCAAT/enhancer binding protein alpha message: transcriptional activator lacking antimitotic activity. Proc Natl Acad Sci U S A1993; 90, 9606-9610.

Liphardt J, Napthine S, Kontos H, Brierley I. Evidence for an RNA pseudoknot loop-helix interaction essential for efficient −1 ribosomal frameshifting. J Mol Biol 1999; 288:321-35.

Lisacek F, Diaz Y, Michel F. J Mol Biol 1994; 235:1206-1217.

Lohse PA, Szostak JW. Ribozyme-catalyzed amino-acid transfer reactions. Nature 1996; 381: 442-444.

Lomakin IB, Hellen CU, Pestova TV. Physical association of eukaryotic initiation factor 4G (eIF4G) with eIF4A strongly enhances binding of eIF4G to the internal ribosomal entry site of encephalomyocarditis virus and is required for internal initiation of translation. Mol Cell Biol 2000; 20: 6019-6029.

Lomeli H, Mosbacher J, Melcher T, Höger T, Geiger JRP, Kuner T, Monyer H, Higuchi M, bach A, Seeburg P. Control of kinetic properties of AMPA receptor channels by nuclear RNA editing. Science 1994; 266: 1709-1713.

Lonergan KM, Gray MW. Editing of transfer RNAs in *Acanthamoeba castellanii* mitochondria. Science 1993; 259: 812-816.

Long DM, Uhlenbeck OC. Self-cleaving catalytic RNA. FASEB J 1993; 7: 25-30.

Lowe TM, Eddy SR. A computational screen for methylation guide snoRNAs in yeast. Science 1999; 283:1168-1171.

Lu XM, Fischman AJ, Jyawook SL, Hendricks K, Tompkins RG, Yarmush ML. Antisense DNA delivery in vivo: liver targeting by receptor-mediated uptake. J Nucl Med 1994;35:269-275.

Lund E, Dahlberg JE. Proofreading and aminoacylation of tRNAs before export from the nucleus [see comments]. Science 1998; 282: 2082-2085.

Luo G, Chao M, Hsieh SY, Sureau C, Nishikura K, Taylor J. A specific base transition occurs on replicating hepatitis delta virus RNA. J Virol 1990; 64: 1021-1027.

Luo Y, Kurz J, MacAfee N, Krause MO. C-myc deregulation during transformation induction: involvement of 7SK RNA. J Cell Biochem 1997; 64:313-327.

Lütcke H. Signal recognition particle (SRP), a ubiquitous initiator of protein translocation. Eur J Biochem 1995 Mar 15; 228(3): 531-535*(50)*.

Lykke-Andersen K. Structural characteristics of the stable RNA introns of archaeal hyperthermophiles and their splicing junctions. J Mol Biol 1994; 243:846-855.

Lyon MF. Pinpointing the center. Nature 1996; 379: 116-117.

Ma Y, Mathews MB. Comparative analysis of the structure and function of adenovirus virus--associated RNAs. J Virol 1993; 67:6605-6617.

Macdonald PM, Struhl G. *Cis*-acting sequences responsible for anterior localization of bicoid mRNA in *Drosophila* embryos. Nature 1988; 336:595-59(8).

Macejak DG, Sarnow P. Internal initiation of translation mediated by the 5' leader of a cellular mRNA [see comments]. Nature 1991; 353: 90-94.

Mackie GA. Ribonuclease E is a 5'-end-dependent endonuclease. Nature 1998; 395:720-723.

Maden BEH. The numerous modified nucleotides in eukaryotic ribosomal RNA. Prog Nucleic Acids Res 1990; 39: 241-303.

Madhani HD, Guthrie C. A novel base-pairing interactiuon between U2 and U6 snRNAs suggests a mechanism for the catalytic activation of the spliceosome. Cell 1992; 71: 803-817

Madhani HD, Bordonné R, Guthrie C. Mutliple roles for U6 snRNA in the splicing pathway. Genes Dev 1990; 4: 2264-2277.

Madhani HD, Guthrie C. Dynamic RNA-RNA interactions in the spliceosome. Annu Rev Genet 1994; 28:1-26.

Mahendran R, Spottswood MS, Ghate A, Ling ML, Jeng K, Miller DL. Editing of the mitochondrial small subunit rRNA in *Physarum polycephalum* [published erratum appears in EMBO J 1994 Mar 15;13(6):1493]. Embo J1994; 13, 232-240.

Mahendran R, Spottswood MR, Miller DL. RNA editing by cytidine insertion in mitochondria of Physarum polycephalum. Nature 1991; 349: 434-438.

Maier RM, Hoch B, Zeltz P, Kössel H. Internal editing of the maize chloroplast ndhA transcript restores codons for conserved amino acids. Plant Cell 1992; 4: 609-616

Malim MH, Bohnlein E, Hauber J, Cullen BR. Functional dissection of the HIV-1 Rev *trans*-activator-derivation of a *trans*-dominant repressor of Rev function. Cell 1989; 58: 205-214.

Malyankar UM, Rittling SR, Coumar A, Denhardt DT. The mitogen-regulated protein/proliferin transcript is degraded in primary mouse embryo fibroblasts but not 3T3 nuclei: altered RNA processing correlates with immortalization. Proc Natl Acad Sci USA 1994; 91: 335-339.

Manley JL. A complex protein assembly catalyzes polyadenylation of mRNA precursors. Curr Opin Genet Dev 1995; 5:222-228.

Manley JL, Proudfoot NJ. RNA 3'ends: Formation and function-meeting review. Genes Dev 1994; 8: 259-264.

Maquat LE . Nonsense-mediated RNA decay in mammalian cells: a splicing-dependent means to down-regulate the levels of mRNAs that prematurely terminate translation. In: Sonenberg N, Hershey JWB, Mathews MB (eds.) Translational control of gene expression 2000: 849-868.

Marcand S, Gilson E, Shore D. A protein-counting mechanism for telomere length regulation in yeast. Science 1997; 275: 986-990.

Marchfelder A, Brennicke A, Binder S. RNA editing is required for efficient excision of tRNA(Phe) from precursors in plant mitochondria. J Biol Chem 1996; 271: 1898-1903.

Markussen FH, Michon AM, Breitwieser W, Ephrussi A. Translational control of *oskar* generates short OSK, the isoform that induces polar granule assembly. Development 1995; 121: 3723-3732.

Marczinke B, Fisher R, Vidakovic M, Bloys AJ, Brierley I. Secondary structure and mutational analysis of the ribosomal frameshift signal of rous sarcoma virus. J Mol Biol 1998; 284:205-225.

Martin F, Schaller A, Eglite S, Schümperli D, Müller B . The gene for histone RNA hairpin binding protein is located on human chromosome 4 and encodes a novel type of RNA binding protein. EMBO J 1997; 16: 769-778.

Mascotti DP, Goessling LS, Rup D, Thach RE. Effects of the ferritin open reading frame on translational induction by iron. Prog Nucleic Acids Res Mol Biol 1996; 55:121-133.

Maxwell ES, Fournier MJ. The small nucleolar RNAs. Annu Rev Biochem 1995; 35:897-934.

McEachern M, Blackburn EH. A conserved sequence motif within the exceptional telomeric sequence of budding yeast. Proc Natl Acad Sci USA 1994; 91: 3453-3457.

McGarry TJ, Lindquist S. The preferential translation of *Drosophila* hsp70 mRNA requires sequences in the untranslated leader. Cell 1985; 42: 903-911.

McGrew LL, Dworkin-Rastl E, Dworkin MB, Richter JD. Poly(A) elongation during *Xenopus* oocyte maturation is required for translational recruitment and is mediated by a short sequence element. Genes Dev 1989; 3: 803-815.

McGuire AM, Hughes JD, Church GM. Conservation of DNA regulatory motifs and discovery of new motifs in microbial genomes. Genome Res 2000; 10:744-757.

McKay DB, Wedekind JE . Small ribozymes. In: Gesteland RF, Cech TR, Atkins JF (eds) The RNA world., 2nd edition 1999; 265-286.

McPheeters DS, Abelson J. Mutational analysis of the yeast U2 snRNA suggests a structural similarity to the catalytic core of group I introns. Cell 1992; 71: 819-831.

Mehldau G, Myers G. A system for pattern matching applications on bioComp. Appl Biosci 1993; 9: 299-314.

Melcher T, Maas S, Herb A, Sprengel R, Seeburg PH, Higuchi M. A mammalian RNA editing enzyme. Nature 1996; 379: 460-464.

Meller VH, Wu KH, Roman G, Kuroda MI, Davis RL. rox1 RNA paints the X chromosome of male *Drosophila* and is regulated by the dosage compensation system. Cell 1997; 88: 445-457.

Melefors Ö. Translational regulation in vivo of the *Drosophila* melanogaster mRNA encoding succinate dehydrogenase iron protein via iron responsive elements. Biochem Biophys Res Commun 1996; 221: 437-441.

Melefors Ö, Hentze MW. Translational regulation by mRNA/protein interactions in eukaryotic cells: ferritin and beyond. BioEssays 1993;15: 85-90.

Melefors Ö, Goosen B, Johansson HE, Stripecke R, Gray NK, Hentze MW. Translational control of 5-aminolevulinate synthase mRNA by iron-responsive elements in erythroid cells. J Biol Chem 1993; 268: 5974-5978.

Mendez R, Hake LE, Andresson T, Littlepage LE, Rudermann JV, Richter JD. Phosphorylation of CPE binding factor by Eg2 regulates translation of c-mos mRNA. Nature 2000; 404:302-307.

Mendez R, Kannenganti GKM, Ryan K, Manley JL, Richter JD. Phosphorylation of CPEB by Eg2 mediates the recruitment of CPSF into an active catoplasmic polyadenylation complex. Mol Cell 2000; 6:1253-1259.

Merrick WC, Hershey, JWB. The pathway and mechanism of eukaryotic protein synthesis. In Translational Control, J Hershey, M Mathews, N Sonenberg, eds (Cold Spring Harbor, NY: Cold Spring Harbor Laboratory Press) 1996, 31-69.

Meyuhas O, Hornstein E. Translational control of TOP mRNAs. In: Sonenberg N, Hershey

JWB, Mathews MB (eds.) Translational control of gene expression 2000; 671-693.

Meyuhas O, Avni D, Shama S. Translational control of ribosomal protein mRNAs in eukayotes. In: Translational control (Hershey JWB, Mathews MB and Sonenberg N) Cold Spring Harbor 1996: 363-388.

Michael WM, Choi M, Drezfuss G. A nuclear export signal in hnRNP A1: a signal-mediated, temperature-dependent nuclear protein export pathway. Cell 1995; 83: 415-422.

Michael WM, Eder PS, Dreyfuss G.The K nuclear shuttling domain: a novel signal for nuclear import and nuclear export in hnRNP K protein. EMBO J 1997; 16:3587-3598.

Michel F, Ferat J-L. Structure and activities of group II introns. Annu Rev Biochem 1995; 64:435-461.

Miller D, Mahendran R, Spottswood M, Costandy H, Wang S, Ling ML, Yang N. Insertional editing in mitochondria of *Physarum*. Semin Cell Biol 1993; 4: 261-266

Miller ED, Plante CA, Kim KH, Brown JW, Hemenway C. Stem-loop structure in the 5' region of potato virus X genome required for plus-strand RNA accumulation. J Mol Biol 1998; 284:591-608.

Mills DR, Kramer FR, Spiegelman S. Complete nucleotide sequence of a replicating RNA molecule. Science 1973; 180:916-927.

Mills DR, Peterson RL, Spiegelman S. An extracellular Darwinian experiment with a self-duplicating nucleic acid molecule. Proc. Natl. Acad. Sci. USA 1967; 58:271-274.

Milosavljevic A, Jurka J. Discovering simple DNA sequences by the algorithm significance method. Comp Appl Biosci 1993; 9:407-411.

Miranda G, Schuppli D, Barrera I, Hauserr C, Sogo JM, Weber H. Recognition of bacteriophage Qß plus strand RNA as a template by Qß replicase: role of RNA interactions mediated by ribosomal proteins S1 and host factor. J Mol Biol 1997; 267:1089-1103.

Mirkin CA, Letsinger RL, Mucic RC, Storhoff JJ. A DNA-based method for rationally assembling nanoparticles into macroscopic materials. Nature 1996; 382:581- 583.

Misra R , Reeves PR. Role of micF in the tolC-mediated regulation of OmpF, a major outer membrane protein of *Escherichia coli* K-12. J Bacteriol 1987; 169: 4722-4730.

Mistelli T. Cell biology of transcription and pre-mRNA splicing: nuclear architecture meets nuclear function. J Cell Sci. 2000 Jun;113 (Pt 11):1841-1849. Review.

Mitchell JR, Cheng J, Collins K. A box H/ACA small nucleolar RNA-like domain at the human telomerase RNA 3' end. Mol Cell Biol 1999 Jan;19:567-576.

Mittermaier A, Varani L, Muhandiram DR, Kay LE, Varani G. Changes in side-chain and backbone dynamics identify determinants of specificity in RNA recognition by human U1A protein. J Mol Biol 1999; 294:967-979

Mize GJ, Ruan H, Low JJ, Morris DR. The inhibitory upstream open reading frame from mammalian S-adenosylmethionine decarboxylase mRNA has a strict sequence specificity in critical positions. J Biol Chem 1998; 273, 32500-32505.

Mizuno T, Chou M-Y, Inouye M. A unique mechanism regulating gene expression: Translational inhibtion by a complementary RNA transcript (micRNA). Proc Natl Acad Sci USA 1984; 81:1966-1970.

Mobarak CD, Anderson KD, Morin M, Beckel-Michener A, Rogers SL, Furneaux H, King P, Perrone-Bizzozero NI. The RNA-binding protein HuD is required for GAP-43 mRNA stability, GAP-43 gene expression, and PKC-dependent neurite outgrowth in PC12 cells. Mol Biol Cell 2000; 11:3191-3203.

Moore MJ, Sharp PA. Evidence for two active sites in the spliceosome provided by stereochemistry of pre-mRNA splicing. Nature 1993; 365: 364-368.

Moore MJ, Query CC, Sharp PA. Splicing of precursors to mRNA by the spliceosome. In: The RNA world Gesteland RF, Atkins JF, Plainview, NY: Cold Spring Harbor Lab. Press.1993: 303-357.

Moras D, Poterszman A. Protein-RNA interactions: Getting into the major groove. Curr Biol 1996; 6:530-532.

Morris DR, Geballe AP. Upstream open reading frames as regulators of mRNA translation. Mol

Cell Biol 2000; 20: 8635-8642.

Morrisey JP, Tollervey D. Yeast snR30 is a small nucleolar RNA required for 18S rRNA synthesis. Mol Cell .Biol 1993; 13: 2469-2477.

Morrisey JP, Tollervey D. Birth of the snoRNPs- the evolution of RNase MRP and the eukaryotic pre-rRNA procesing sytem. Trends Biochem Sci 1995; 20: 78-82.

Moss EG, Lee RC, Ambros V. The cold shock domain proten lin-28 contols developmental timing in *C. elegans* and is regulated by the lin-4 RNA. Cell 1997; 88: 637-646.

Mount SM. AT-AC introns: An ATtACk on dogma. Science 1996; 271: 1690-1692.

Mourrain P, Beclin C, Elmayan T, Feuerbach F, Godon C, Morel JB, Jouette D, Lacombe AM, Nikic S, Picault N, Remoue K, Sanial M, Vo TA, Vaucheret H. *Arabidopsis* SGS2 and SGS3 genes are required for posttranscriptional gene silencing and natural virus resistance. Cell 2000; 101:533-542.

Mowry KL, Steitz JA . Identification of the human U7snRNP as one of several factors involved in the 3' end maturation of histone pre-messenger RNAs. Science 1987; 238:1682-1687.

Moxham CM, Malbon CC. Insulin action impaired by deficiency of the G-protein subunit G_{i_2}. Nature 1996;379:840-844.

Moxham CM, Hod Y, Malbon CC. Induction of G alpha i2-specific antisense RNA in vivo inhibits neonatal growth. Science 1993; 260:991-995.

Muckenthaler M, Gray NK, Hentze MW. IRP-1 binding to ferritin mRNA prevents the recruitment of the small ribosomal subunit by the cap-binding complex eIF4F. Mol Cell 1998: 2, 383-388.

Muhlrad D, Parker R. Mutations affecting stability and deadenylation of the yeast MFA2 transcript. Genes Dev 1992; 6:2100-2111.

Muller B, Link J, Smythe C. Assembly of U7 small nuclear ribonucleoprotein particle and histone RNA 3' processing in Xenopus egg extracts. J Biol Chem 2000;275:24284-93.

Müller PP, Hinnebusch AG. Multiple upstream AUG codons mediate translational control of GCN4. Cell 1986; 45: 201-207.

Müllner EW, Kühn LC. A stem-loop in the 3' untranslated region mediates iron-dependent regulation of transferrin receptor mRNA stability in the cytoplasm. Cell 1988; 53:815-825.

Murray JB, Terwey DP, Maloney L, Karpeisky A, Usman N, Beigelman L, Scott WG. The structural basis of hammerhaed ribozyme self-cleavage. Cell 1998; 92: 665-673.

Musco G, Stier G, Joseph C, Castiglione-Morelli MA, Nilges M, Gibson TJ, Pastore A. Three-dimensional structure and stability of the KH domain: molecular insights into the fragile X syndrome. Cell 1996; 85:237-245.

Nagai K, Mattaj IW. RNA-protein interactions. IRL Press, Oxford, N.Y., 1994.

Nagalla SR, Barry BJ, Spindel ER. Cloning of complementary DNAs encoding the amphibian bombesin-like peptides Phe8 and Leu8 phyllolitorin from *Phyllomedusa sauvagei*: potential role of U to C RNA editing in generating neuropeptide diversity. Mol Endocrinol 1994; 8: 943-951.

Nakamura TM et al. Telomerase catalytic subunit homologs from fission yeast and human. Science 1997; 277:955-959.

Nakielny S, Dreyfuss G. Nuclear export of proteins and RNAs. Curr Op Cell Biol 1997;9:420-429.

Narayanam R, Akhtar S. Antisense therapy. Curr Opin Oncol 1996; 8:509-515.

Navaratnam N, Patel D, Shah RR, Greeve JC, Powell LM, Knott TJ, Scott J. An additional editing site is present in apolipoprotein B mRNA. Nucleic Acids Res 1991; 19: 1741-1744.

Nedde DN, Ward MO. Visualizing relationships between nucleic acid sequences using correlation images. Comp Appl Biosci 1993; 9: 331-335.

Nesbitt SM, Erlacher HA, Fedor MJ. The internal equilibrium of the hairpin ribozyme: temperature, ion and pH effects. J Mol Biol 1999; 286:1009-1024.

Neugebauer KM, Roth MB. Distribution of pre-mRNA splicing factors at sites of RNA polymerase II transcription. Genes Dev 1997; 11:1148-1159.

Newman A. Small nuclear RNAs and pre-mRNA splicing. Curr Op in Cell Biol 1994; 6:360-

367.

Newmann A, Norman C. U5 snRNAs interacts with exon sequences at 5' and 3' splice sites. Cell 1992; 68: 743-754.

Ni J, Tien, A, Fournier M. Small nucleolar RNAs direct site-specific synthesis of pseudouridine in ribosomal RNA. Cell 1997; 89: 565-573.

Nichols RC, Wang XW, Tang J, Hamilton BJ, High FA, Herschan HR, Rigby WF. The RGG domain in hnRNP A2 affects subcellular localization. Exp Cell Res 2000; 256: 522-532.

Nicoll M, Akerib CC, Meyer BJ. X-chromosome-counting mechanisms that determine nematode sex. Nature 1997; 388: 200-204.

Nicoloso M, Qu LH, Michot B, Bachellerie J-P. Intron-encoded, antisens small nucleolar RNAs: THe characterization of nine novel species points to their direct role as guides for the 2'-O-ribose methylation of rRNAs. J Mol Biol 260: 178-195.

Nilsen TW. RNA-RNA interactions in the spliceosome: unravelling the ties that bind. Cell 1994a; 78: 1-4.

Nilsen TW. Unusual strategies of gene expression and control in parasites. Science 1994b; 264, 1868-1869.

Nilsen TW. Trans-splicing: an update. Mol Biochem Parasitol 1995; 73: 1-6.

Nilsen TW. A parallel spliceosome. Science 1996; 273: 1813

Novick RP, Ross HF, Projan SJ et al. Synthesis of staphylococcal virulence factors is controlled by a regulatory RNA molecule. EMBO J 1993; 12:3967-3975.

Nowakowski J, Tinoco I. RNA structure and stability. Semina in Virol; 1997; 8:153-165.

Nugent JM, Palmer JD. RNA-mediated transfer of the gene coxII from the mitochondrion to the nucleus during flowering plant evolution. Cell 1991; 66: 473-481.

Nyce JW, Metzger WJ. DNA antisense therapy for asthma in an animal model. Nature 1997; 385:721-725.

Ogura H,Agata H, Xie M, Odaka T, Furutani H. A study of learning splice sites of DNA sequence by neural networks. Comput Biol Med 1997; 27:67-75.

Oh SK, Scott MP, Sarnow P. Homeotic gene Antennapedia mRNA contains 5'-noncoding sequences that confer translational initiation by internal ribosome binding. Genes Dev 1992; 6: 1643-1653.

Ohno M, Ségref A, Bachi A, Wilm M, Mattaj IW. PHAX, a mediator of U snRNA nuclear export whose activity is regulated by phosphorylation. Cell 2000; 101:187-198.

Oliver AW, Bogdarina I, Schroeder E, Taylor IA, Kneale GG. Preferential binding of fd gene 5 protein to tetraplex nucleic acid structures. J Mol Biol 2000; 301:575-584.

Olsthoorn RC, Licis N, Van Duin J. Leeway and constraints in the forced evolution of a regulatory RNA helix. EMBO J 1994; 13:2660-2668.

Olsthoorn RC, Garde G, Dayhuff T, Atkins JF, Van Duin J. Nucleotide sequence of a single stranded RNA phage from *Pseudomonas aeruginosa*: kinship to coliphages and conservation of regulatory RNA structures. Virology 1995; 206:611-625.

Omer AD, Lowe TM, Russell AG, Ebhardt H, Eddy SR, Dennis PP. Homologs of small nucleolar RNAs in Archaea. Science 2000; 288:517-22.

Osada Y, Saito R, Tomita M. Analysis of base-pairing potentials between 16S rRNA and 5' UTR for translation initiation in various prokaryotes. Bioinformatics. 1999 Jul-Aug;15(7-8):578-581.

Oskouian B, Rangan VS, Smith S. Regulatory elements in the first intron of the rat fatty acid synthase gene. Biochem J 1997; 324:113-121.

Ossipow V, Descombes P, Schibler U. CCAAT/enhancer-binding protein mRNA is translated into multiple proteins with different transcription activation potentials. Proc Natl Acad Sci USA 1993; 90: 8219-8223.

Ostareck DH, Ostareck-Lederer A, Wilm M, Thiele BJ, Mann M, Hentze MW. mRNA silencing in erytroid differentiation: hnRNP K and hnRNP E1 regulate 15-lipoxygenase translation from the 3' end. Cell 1997; 89:597-606.

Ostareck DH, Ostareck-Lederer A, Shatsky IN, Hentze MW . Lipoxygenase mRNA silencing in erythroid differentiation: The 3'UTR regultaory complex controls 60S ribosomal subunit joining. Cell 2001, in press.

Ostareck-Lederer A, Ostareck DH, Standart N, Thiele BJ. Translation of 15-lipoxygenase mRNA is inhibited by a protein that binds to a repeated sequence in the 3' untranslated region. EMBO J 1994; 13:1476-1481.

Ostareck-Lederer A, Ostareck DH, Hentze MW. Cytoplasmic regulatory functions of the KH-domain proteins hnRNPs K and E1/E2. Trends Biochem Sci 1998; 23, 409-411.

Pachnis V, BelayewA, Tilghman S. Locus unlinked to _-fetoprotein under the control of the murine raf and Rif genes. Proc Natl Acad Sci USA 1984; 81: 5523-5527.

Paillart J-C, Berthoux L, Ottmann M, Darlix J-L, Marquet R, Ehresmann B, Ehresmann C. A dual role of the putative RNA dimerization initiation site of human immunodeficiency virus type 1 in genomic RNA packaging and proviral DNA synthesis. J Virol 1996; 8348-8354.

Palla F, Melfi R, Di Geatano L, Bonura C, Anello L, Alessandro C, Spinelli G. Regulation of the sea urchin early H2A histone gene expression depends on the modulator element and on sequences located near the 3' end. Biol Chem 1999; 380:159-165.

Pan T. Novel RNA substrates for the ribozyme from *Bacillus subtilis* ribonuclease P identified by in vitro selection. Biochemistry 1995; 34: 8458-8464.

Pan T, Dichtl B, Uhlenbeck O. Properties of an in vitro selected Pb++ cleavage motif. Biochemistry 1994; 33: 9561-9565.

Pandey NB, Marzluff WF. The stem-loop structure at the 3'end of histone mRNA is necesary and sufficient for regulation of histone mRNA stability. Mol Cell Biol 1987; 7:4557-4559.

Parker R, Siliciano PG, Guthrie C. Recognition of the TACTAAC box during mRNA splicing in yeast involves base pairing to the U2-like snRNA. Cell 1987; 49: 229-239

Parker R, Simmons T, Shuster EO, Siliciano PG, Guthrie C. Genetic analysis of small nuclear RNAs *in Saccharomyces cerevisiae*: viable sextuple mutant. Mol Cell Biol 1988; 8: 3150-3159.

Paul AV, van Boom JH, Filippov D, Wimmer E. Protein-primed RNA synthesis by purified poliovirus RNA polymerase. Nature 1998; 393:280-284.

Pedersen AG, Baldi P, Chauvin Y, Brunak S. DNA structure in human RNA polymerase II promoters. J Mol Biol 1998; 281:663-673.

Pelletier J, Kaplan G, Racaniello VR and Sonenberg N. Cap-independent translation of poliovirus mRNA is conferred by sequence elements within the 5' noncoding region. Mol Cell Biol 1988; 8: 1103-1112.

Pellizzoni L, Lotti F, Rutjes SA, Pierandrei-Amaldi P. Involvement of the Xenopus laevis Ro60 autoantigen in the alternative interaction of La and CNBP proteins with the 5'UTR of L4 ribosomal protein mRNA. J Mol Biol 1998; 281:593-608.

Peltz SW, Ross J. Autogenous rgulation of histone mRNA decay by histone proteins in a cell-free system. Mol Cell Biol 1987;7:536-540.

Peltz SW, Brown AH and Jacobson A. mRNA destabilization triggered by premature translational termination depends on at least three cis-acting sequence elements and one trans-acting factor. Genes Dev 1993; 7, 1737-1754.

Pelle R, Murphy NB. In vivo UV-cross-linking hybridization: a powerful technique for isolating RNA binding proteins. Application to trypanosome mini-exon derived RNA. Nucleic Acids Res 1993; 25: 2453-2458.

Pellizzoni L, Lotti F, Maras B, Pierandrei-Amaldi P. Cellular nucleic acid binding protein binds a conserved region of the 5'UTR of *Xenopus laevis* ribosomal protein mRNA.1997; J Mol Biol 1997;267:264-275.

Penalva LO, Ruiz MF, Ortega A, Granadino B, Vincente L, Segarra C, Valcarcel J, Sanchez L. The Drosophile fl(2)d gene, required for female-specific splicing of Sxl and tra pre-mRNAs, encodes a novel nuclear protein with HQ-rich domain. Genetics 2000; 155:129-139.

Penny GD, Kay GF, Sheardown SA, Rastan S, Brockdorff N. Requirement for XIST in X chromosome inactivation. Nature 1996; 379: 131-137;

Percudani R, Pavesi A, Ottonello S. Transfer RNA gene redundancy and translational selection in Saccharomyces cereviaise. J.Mol Biol 1997; 268, 322-330.

Perry RP. RNA processing comes of age. J Cell Biol 1981; 91:28s-38s.

Pesole G, Liuni S, D'Souza M. PatSearch: a pattern matcher software that finds functional elements in nucleotide and protein sequences and assesses their statistical significance. Bioinformatics 2000 May;16:439-450.

Pestova TV, Hellen CU, Wimmer E. A conserved AUG triplet in the 5' nontranslated region of poliovirus can function as an initiation codon in vitro and in vivo. Virology 1994; 204: 729-737.

Pestova TV, Hellen CU, Shatsky IN. Canonical eukaryotic initiation factors determine initiation of translation by internal ribosomal entry. Mol Cell Biol 1996; 16: 6859-6869.

Pestova TV, Lomakin IB, Lee JH, Choi SK, Dever TE, and Hellen CU (2000). The joining of ribosomal subunits in eukaryotes requires eIF5B. Nature (403), 332-5.

Pestova TV, Shatsky IN, Fletcher SP, Jackson RJ, Hellen CU. A prokaryotic-like mode of cytoplasmic eukaryotic ribosome binding to the initiation codon during internal translation initiation of hepatitis C and classical swine fever virus RNAs. Genes Dev 1998;12:67-83.

Pokrywka NJ, Stephenson EC. Microtubules are a general component of mRNA localization systems in *Drosophila* oocytes. Dev Biol 1995; 167: 363-370.

Polson AG, Bass BL. Preferential selection of adenosines for modification by double-stranded RNA adenosine deaminase. EMBO J 1994; 13: 5701-5711.

Polson AG, Bass BL, Casey JL. RNA editing of hepatitis delta virus antigenome by dsRNA-adenosine deaminase. Nature 1996; 380: 454-456.

Poola I, Koduri S, Chatra S, Clarke R. Identification of twenty alternatively spliced estrogen receptor alpha mRNAs in breast cancer cell lines and tumors using splice targeted primer approach. J Steroid Biochem Mol Biol 2000; 72:249-58.

Powell LM, Wallis SC, Pease RJ, Edwards YH, Knott TJ, Scott J. A novel form of tissue-specific RNA processing produces apolipoprotein-B48 in intestine. Cell 1987; 50: 831-840.

Powers T, Noller HF. The 530 loop of 16S rRNA: a signal to EF-Tu? Trends Genet 1994; 10, 27-31.

Preiss T, Hentze MW. Dual function of the messenger RNA cap structure in poly(A)-tail-promoted translation in yeast. Nature 1998; 392:516-20.

Prescutti C, Ciafr SA, Bozzoni I. The ribosomal protein L2 in *S.cerevisiae* controls the level of accumulation of its own mRNA. EMBO J 1991; 8:2215-2221.

Price DH and Gray MW. A novel nucleotide incorporation activity implicated in the editing of mitochondrial transfer RNAs in Acanthamoeba castellanii. Rna 1999; 5, 302-317.

Price DH, Gray MW . Editing of tRNA. In: Grosjean H and Benne R (eds.). Modification and Editing of RNA 1998; 289-305.

Price SR, Evans PR, Nagai K. Crystal structure of the spliceosomal U2B"-U2A' protein complex bound to a fragment of U2 small nuclear RNA. Nature 1998; 394:645-650.

Proud CG. p70 S6 kinase: an enigma with variations. Trends Biochem 1996; 23:181-184.

Puoti A, Kimble J. The hermaphrodite sperm/oocyte switch requires the *Caenorhabditis elegans* homologs of PRP2 and PRP22. Proc Natl Acad Sci USA 2000; 97:3276-3281.

Purdey M. The UK epidemic of BSE: slow virus or chronic pestizide initiated modification of the prion protein ? Med Hypothesis 1996; 46:445-454.

Pütz J, Florentz C, Benselet F, Giegé R. A single methyl group prevents the mischarging of a tRNA. Struct Biol 1994; 1: 580-582.

Pyronnet S, Pradayrol L, Sonenberg N. A cell cycle-dependent internal ribosome entry site. Mol Cell 2000; 5, 607-616.

Raghunathan,PL, Guthrie,C. RNA unwinding in U4/U6 snRNPs requires ATP hydrolysis and the DEIH-box splicing factor Brr2. Curr Biol 1998; 8:847-855

Ralle T, Gremmels D, Stick R. Translational control of nuclear lamin B1 mRNA during oogenesis and early development of Xenopus. Mech Dev 1999; 84:89-101.

Ramchandani S, MacLeod AR, Pinard M, von Hofe E, Szyf M. Inhibition of tumorigenesis by a cytosine-DNA, methyltransferase, antisense oligodeoxynucleotide. Proc Natl Acad Sci USA

1997; 94:684-689.

Raney A, Baron AC, Mize GJ, Law GL, Morris DR. In vitro translation of the upstream open reading frame in the mammalian mRNA encoding S-adenosylmethionine decarboxylase. J Biol Chem 2000; 275: 24444-24450.

Rastan S. Non-random X-chromosome inactivation in mouse X-autosome translocation embryos: location of the inactivation centre. Embryol Exp Morphol 1983; 78: 1-22.

Rastan S, Brown SDM. The search for the mouse X-chromosome inactivation centre. Genet Res 1990; 56: 99-106.

Ratajczak MZ, Kant JA, Luger SM et al. . In vivo treatment of human leukemia in a scid mouse model with c-myb antisense oligodeoxynucleotides. Proc Natl Acad Sci USA 1992; 89:11823-11827.

Reich CI, VanHoy RW, Porter GL, Wise JA. Mutations at the 3' splice site can be supressed by compensatory base changes in U1 snRNA in fission yeast. Cell 1992; 69: 1159-1169.

Reinhart BJ, Slack FJ, Basson M, Pasquinelli AE, Bettinger JC, Rougvie AE, Horvitz HR, Ruvkun G. The 21-nucleotide let-7 RNA regulates developmental timing in *Caenorhabditis elegans*. Nature 2000; 403:901-906.

Rendahl KG, Jones KR, Kulkarni SJ, Bagully SH, Hall JC. The dissonance at the no-on-transient-A locus of *D.melanogaster*: genetic control of courtship song and visual behaviors by a protein with putative RNA binding domain. J Neurosci 1992; 12:390-407.

Ribes V, Römisch K, Giner A, Dobberstein B, Tollervey D. *E.coli* 4.5S RNA is part of a ribonucleoprotein particle that has properties related to signal recognition particle. Cell 1990; 63:591-600.

Richter JD. Translational control during early development. BioEssays 1991; 13:179-183.

Richter JD. Cytoplasmic polyadenylation in development and beyond. Microbiol. Mol. Biol. Rev. 1999; 63:446-456.

Richter JD. Influence of polyadenylation-induced translation on metatoan development and neuronal synaptic function. In: Sonenberg N, Hershey JWB, Mathews MB (eds.) Translational control of gene Expression. Cold Spring Harbor Laboratory Press, Cold Spring Harbor, New York 1996:785-805.

Rinke J, Appel B, Digweed M, Luhrmann R. Localization of a base-paired interaction between small nuclear RNAs U4 and U6 in intact U4/U6 ribonucleoprotein particles by psoralen cross-linking. J Mol Biol 1985; 185: 721-731

Rio DC. Splicing of mRNA, modulation, regulation and role in development. Curr Opin Genet Dev 1993; 3: 574-584.

Ripmaster TL, Woolford JL Jr. A protein containing conserved RNA-recognition motifs is associated with the ribosomal subunits in *Saccharomyces cerevisiae*. Nucleic Acids Res 1993; 21:3211-3216.

Rizzetto M, Canese MG, Gerin JL, London WT, Sly DL, Purcell RH. Transmission of the hepatitis B virus-associated delta antigen to chimpanzees. J Infect Dis 1980a; 141: 590-602

Rizzetto M, Hoyer B, Canese MG, Shih JWK, Purcell RH, Gerin JL. delta Agent: association of delta antigen with hepatitis B surface antigen and RNA in serum of delta-infected chimpanzees. Proc Natl Acad Sci USA 1980b; 77: 6124-6128.

Robbins J, Dilworth SM, Laskey RA, Dingwall C. Two interdependent basic domains in nucleoplasmin nuclear targeting sequence: identification of a class of bipartite nuclear targeting sequences. Cell 1991; 64:615-623.

Rogers J, Wall R. A mechanism for RNA splicing. Proc Natl Acad Sci USA 1980; 77: 1877-1879.

Rother M, Wilting R, Commans S, Bock A. Indentification and characterisation of the selenocysteine-specific translation factor SelB from the archaeon *Methanococcus jannaschii*. J Mol Biol 2000; 299:351-358.

Rouault TA, Hentze MW, Dancis A, Caughman W, Harford JB, Klausner RD. Influence of altered transcription on the translational control of human ferritin expression. Proc Natl Acad Sci USA 1987; 84: 6335-6339.

Rouault TA, Hentze MW, Haile DJ et al. The iron-responsive element binding protein: A method for the affinity purification of a regulatory RNA-binding protein. Proc Natl Acad Sci USA 1989; 86: 5768-5772.

Rueter SM, Burns CM, Coode SA, Mookherjee P; Emeson RB. Glutamate receptor RNA editing in vitro by enzymatic conversion of adenosine to inosine. Science 1995; 267:1491-1494.

Rueter SM, Dawson TR, Emeson RB. Regulation of alternative splicing by RNA editing. Nature 1999; 399:75-80.

Ruiz-Echevarria MJ, Peltz SW. The RNA binding protein Pub1 modulates the stability of transcripts containing upstream open reading frames. Cell 2000; 101:741-751.

Ruskin B, Krainer AR, Maniatis T, Green MR. Excision of an intact intron as a novel lariat structure during pre-mRNA splicing in vitro. Cell 1984; 38: 317-331.

Ruvkun G, Giusto J. The Caenorhabditis elegans heterochronic gene lin-14 encodes a nuclear protein that forms a temporal developmental switch. Nature 1989; 338: 313-319

Ruvolo V, Altszuler R, Levitt A. The transcript encoding the circumsporozoite antigen of *plasmodium berghei* utilizes heterogeneous polyadenylation sites. Mol Biochem Parasitol 1993; 57: 137-150.

Ryder SP, Strobel SA. Nucleotide analog interference mapping of the hairpin ribozyme: implications for secondary and tertiary structure formation. J Mol Biol 1999; 291:295-311.

Rymond BC, Rosbash M. Yeast pre-mRNA splicing. Volume II. The molecular and cellular biology of the yeast *Saccharomyces*: gene expression. Cold Spring Harbor Laboratory Press, Cold Spring Harbor 1992:143-163.

Sachs A . Physical and functional interactions between the mRNA cap structure and the poly(A) tail. In: Sonenberg N, Hershey JWB, Mathews MB (eds) Translational control of gene expression. Cold Spring Harbor Laboratory Press, Cold Spring Harbor, New York 2000:447-465.

Sachs AB, Kornberg RD . Nuclear polyadenylate binding protein. Mol Cell Biol 1985; 5:1993-1996.

Sachs AB, Sarnow P, Matthias MW. Starting at the beginning, middle, and end: Translation initiation in eukaryotes. Cell 1997;89:831-838.

Saenger W . Principles of nucleic acid structure. Springer Berlin Heidelberg New York 1984.

Saks ME, Sampson JR, Abelson J. Evolution of a transfer RNA gene through a point mutation in the anticodon. Science 1998; 279:1665-1670.

Saldanha R, Mohr G, Belfort M, Lambowitz AM. Group I and group II introns. FASEB J 1993; 7:15-24.

Sanchez L, Granadino B, Torres M. Sex determination in *Drosophila* melanogaster: X-linked genes involved in the initial step of sex-lethal activation. Dev Genet 1994; 15: 251-264.

Sankaranarayanan R, Dock-Bregeon AC, Romby P, Caillet J, Springer M, Rees B, Ehresmann C, Ehresmann B, Moras D. The structure of threonyl-tRNA synthetase-tRNA(Thr) complex enlightens its repressor activity and reveals an essential zinc ion in the active site. Cell 1999; 97:371-81.

Sarver N, Cairns S. Ribozyme trans-splicing and RNA tagging: Following the messenger. Nat Med 1996; 2: 641-642.

Sasaki H, Jones PA, Chaillet JR, Ferguson-Smith AC, Barton S, Reik W, Surani A. Parental imprinting: potentially active chromatin of the repressed maternal allele of the mouse insulin-like growth factor II (Igf2) gene. Genes Dev 1992; 6: 1843-1856.

Scarabino D, Tocchini-Valentini GP. Influence of substrate structure on cleavage by hammerhead ribozyme. FEBS Lett. 1996; 383:185-90.

Scherly D, Boelens W, Dathan NA, van Venrooij WJ, Mattaj IW. Major determinants of the specificity of interaction between small nuclear ribonucleoproteins U1A and U2B" and their cognate RNAs. Nature 1990; 345:502-506.

Schmidt-Zachmann MS, Nig EA. Protein localization to the nucleolus: A search for targeting domains in nucleolin. J Cell Sci 1993; 799-806.

Schmitt ME. Molecular modeling of the three-dimensional architecture of the RNA component of yeast RNase MRP. J Mol Biol 1999; 292:827-836.

Schultes EA, Bartel DP. One sequence, two ribozymes: implications for the emergence of new ribozyme folds. Science 2000; 289:448-452.

Schuppli D, Miranda G, Qiu S, Weber H. A branched stem-loop structure in the M-site of bacteriophage Qbeta RNA is important for template recognition by Qbeta replicase holoenzyme. J Mol Biol 1998; 283:585-593.

Schuster P, Stadler PF, Renner A. RNA structure and folding: from conventional to new issues in structure prediction. Curr Opin Struct Biol 1997; 7:229-235.

Schuster W, Brennicke A. Plastid, nuclear and reverse transcriptase sequences in the mitochondrial genome of Oenothera: is genetic information transferred between organelles via RNA? EMBO J 1987; 6: 2857-2863.

Schuster W, Hiesel R, Brennicke A. RNA editing in plant mitochondria. Semin Cell Biol 1993; 4: 279-284.

Schwartz DC, Parker R. mRNA decapping in yeast requires dissociation of the cap binding protein, eukaryotic translation initiation factor 4E. Mol Cell Biol 2000; 20:7933-7942.

Scott J. A place in the world for RNA editing. Cell 1995; 81:833-836.

Scott J. Messenger RNA editing and modification. Curr Opin Cell Biol 1989; 1: 1141-1147.

Searls DB. Doing sequence analysis with your printer. Comp Appl Biosci 1993; 9: 421-426.

Ségref A, Sharma K, Doye V, Hellwig A, Huber J, Lührmann R, Hurt E. Mex67p, a novel factor for nuclear mRNA export, binds to both poly(A)$^+$ RNA and nuclear pores. EMBO J 1997; 16: 3256-3271.

SenGupta DJ, Zhang B, Kraemer B,Pochart P,Fields S, Wickens M. A three-hybrid system to detect RNA-protein interactions in vivo. Proc Natl Acad Sci USA 1996; 93:8496-8501.

Serano TL, Cohen RS. A small predicted stem-loop structure mediates oocyte localization of *Drosophila* K10 mRNA. Development 1995; 121: 3809-3818.

Seraphin B. How many intronic RNAs? Trends Biochem Sci 1993; 18:330-331.

Service RF. New probes open windows on gene expression, and more. Science 1998; 280:1010-1011.

Seto AG, Zaug AJ, Sobel SG, Wolin SL, Cech TR. *Saccharomyces cerevisiae* telomerase is an Sm small nuclear ribonucleoprotein particle. Nature 1999; 401:177-280.

Shah SA, Brunger AT. The 1.8 A crystal structure of a statically disordered 17 base-pair RNA duplex: principles of RNA crystal packing and its effect on nucleic acid structure. J Mol Biol 1999; 285:1577-1588.

Sharma PM, Bowman M, Madden SL, Rauscher FJ 3rd, Sukumar S. RNA editing in the Wilms' tumor susceptibility gene, WT1. Genes Dev 1994; 8:720-731.

Sharp PA. Trans splicing: variations on a familiar theme? Cell 1987; 50: 147-148

Sharp PA. RNAi and double-strand RNA. Genes Dev. 1999; 13:139-141.

Shatkin AJ . Capping of eukaryotic mRNAs Cell 1976; 9:645-653.

Shaw G, Kamen R. A conserved AU sequence from the 3'untranslated region of GM-CSF mRNA mediates selective mRNA degradation. Cell 1986; 46: 659-667.

Shen Q, Chu FF, Newburger PE. Sequences in the 3'untranslated region of the human cellular glutathione peroxidase gene are necessary and sufficient for selenocysteine incorporation at the UGA Codon. J Biol Chem 1993; 268:11463-11469.

Shiman R, Draper DE. Stabilization of RNA tertiary structure by monovalent cations. J Mol Biol 2000; 302:79-91.

Shimizu A. Molecular mechanisms for immunoglobulin class switching and IgE production. Nippon Rinsho 1996; 54: 440-445.

Short S, Tian D, Short ML, Jungmann RA. Structural determinants for posttranscriptional stabilization of lactate dehydrogenase A mRNA by the protein kinase C signal pathway. J Biol Chem 2000; 275:12963-9.

Shyu AB, Greenber ME, Belasco JG. The c-fos transcript is targeted for rapid decay by two distinct mRNA degradation pathways. Genes Dev 1989; 3:60-72.

Shyu AB, Belasco J, Greenberg ME. Two distinct destabilizing elements in the c-fos message trigger deadenylation as a first step in rapid mRNA decay. Genes Dev 1991; 5: 221-231.

Sibbald PR, Sommerfeld H, Argos P. Overseer: a nucleotide sequence searching tool. Comp Appl Biosci 1992; 8:45-48.

Siegel V, Walter P. Removal of the Alu structural domain from signal recognation particle leaves its protein translocation activity intact. Nature. 1986; 320:81-84.

Sierakowska H , Sambade MJ, Agrawal S, Kole R. Repair of thalassemic human ß-globin mRNA in mammalian cells by antisense oligonucleotides. Proc Natl Acad Sci USA 1996; 93:12840-12844.

Simard MJ, Chabot B. Control of hnRNP A1 alternative splicing: an intron element represses use of the common 3'splice site. Mol Cell Biol 2000; 20:7353-7562.

Simons RW, Kleckner N. Translational control of IS10 transposition. Cell 1983; 34: 683-691.

Simons M, Edelman ER, DeKeyser JL, Langer R, Rosenberg RD. Antisense c-myb oligonucleotides inhibit intimal arterial smooth muscle cell accumulation in vivo. Nature 1992; 359:67-70.

Simpson L . RNA editing- An evolutionary perspective. In: Gesteland RF, Cech TR, Atkins JF (eds.) The RNA World 2nd edn 1999:585-608.

Simpson L, Shaw J. RNA editing and the mitochondrial cryptogenes of kinetoplastid protozoa. Cell 1989; 57: 355-366.

Simpson L, Thiemann OH. Sense from nonsense: RNA editing in mitochondria of kinetoplastid protozoa and slime molds. Cell 1995; 81:837-840.

Singer MS, Gottschling DE . TLC1: template RNA component of Saccharomyces cerevisiae telomerase. Science 1994; 266: 404-409.

Siomi H, Dreyfuss G. RNA-binding proteins as regulators of gene expression. Curr Opin Genet Dev 1997; 7:345-353.

Sit TL, Vaewhongs AA, Lommel SA. RNA-mediated trans-activation of transcription from a viral RNA. Science 1998; 281:829-832.

Skuse GR, Cappione AJ, Sowden M, Metheny LJ and Smith HC. The neurofibromatosis type I messenger RNA undergoes base-modification RNA editing. Nucleic Acids Res 1996; 24, 478-485.

Smith HC. Apolipoprotein B mRNA editing: the sequence to the event. Semin Cell Biol 1993; 4: 267-278

Smith S, de Lange T. TRF1, a mammalian telomeric protein. Trends Genet 1997; 13: 21-26.

Solovyev VV, Lawrence CB. Identifiaction of human gene functional regions based on oligonucleotide composition. Ismb 1993; 1:371-379.

Soma A, Kumagai, Nishikawa K, Himeno H. The anticodon loop is a major determinant of *Saccharomyces cerevisiae* tRNA. J Mol Biol 1996; 263:707-714.

Sontheimer EJ, Steitz JA. The U5 and U6 small nuclear RNAs as active site components of the spliceosome. Science 1993; 262: 1989-1996

Srivastava SP, Davies MV, Kaufman RJ. Calcium depletion from the endoplasmic reticulum activates the double-stranded RNA-dependent protein kinase(PKR) to inhibit protein synthesis. J Biol Chem 1995; 270: 16619-16624.

Staden R. methods for discovering novel motifs in nucleic acid sequences. Comp Appl Biosci 1989; 5: 293-298.

Staden R. The Staden sequence analysis package. Mol Biotechnol 1996; 5: 233-241.

Staden R, Beal KF, Bonfield JK. The Staden package. Methods Mol Biol 2000; 132:115-130.

Stams T, Niranjanakumari S, Fierke CA, Christianson DW. Ribonuclease P protein structure: evolutionary origins in the translational apparatus. Science 1998; 280:752-755.

Stebbins-Boaz B, Richter JD. Translational control during early development. Crit Rev Eukaryot Gene Expr 1997; 7:73-94.

Stebbins-Boaz B, Cao Q, de Moor CH, Mendez R, Richter JD. Maskin is a CPEB-associated factor that transiently interacts with eIF4E. Mol Cell 1999; 4:1017-1027.

Steinhauser S, Beckert S, Capesius I, Malek O, Knoop V. Plant mitochondrial RNA editing. J Mol Evol 1999; 48:303-312.

Steinmann-Zwicky M. Sex deteermination of the *Drosophila* germ line: tra and dsx control somatic inductive signals. Development 1994; 120: 707-716.

Steitz JA. Splicing takes a holliday. Science 1992; 257: 888-889.

Stöger R, Kubicka P, Liu CG, Kafri T, Razin A, Cedar H, Barlow DP. Maternal-specific methylation of the imprinted mouse Igf2r locus identifies the expressed locus as carrying the imprinting signal. Cell 1993; 73: 61-71.

Stuart K. The RNA editing process in trypanosoma brucei. Semin Cell Biol 1993; 4: 251-260.

Stuart K. RNA editing in mitochondrial mRNA of trypanosomatids. Trends Biochem Sci 1991; 16: 68-72.

Stuart K, Allen TE, Heidmann S, Seiwert SD. RNA editing in kinetoplastid protozoa. Microbiol Mol Biol Rev 1997; 61, 105-120.

Sturm NR, Simpson L. Kinetoplastid DNA minicircles encode guide RNAs for the editing of cytochrome oxidase subunit III mRNA. Cell 1990; 61: 879-884.

Sudarsanakumar C, Xiong Y, Sundaralingam M. Crystal structure of an adenine bulge in the RNA chain of a DNA.RNA hybrid, d(CTCCTCTTC).r(gaagagagag). J Mol Biol 2000; 299:103-12.

Sullenger BA, Cech TR. Tethering ribozymes to a retroviral packaging signal for destruction of viral RNA. Science 1993; 262: 1566-1569.

Sullenger BA, Cech TR. Ribozyme-mediated repair of defective mRNA by targed trans-splicing. Nature 1994; 371: 619-622.

Surdej P, Riedl A, Jacobs-Lorena M. Regulation of mRNA stability in development. Annu Rev Genet. 1994; 28:263-282.

Sutcliffe JS, Nakao M, Christian S, Orstavik KH, Tommerup N, Ledbetter DH, Beaudet AL. Deletions of a differentially methylated CpG island at the SNRPN gene define a putative imprinting control region. Nat Genet 1994; 8:52-58.

Swanson MS, Dreyfuss G. Classification and purification of proteins of heterogeneous nuclear ribonucleoprotein particles by RNA-binding specificities. Mol Cell Biol 1988; 8:2237-2241.

Symons RH. Small catalytic RNAs. Annu Rev Biochem 1992; 61:641-671.

Tabara H, Sarkissian M, Kelly WG, Fleenor J, Grishok A, Timmons L, Fire A, Mello CC. The rde-1 gene, RNA interference, and transposon silencing in *C. elegans*. Cell 1999; 99:123-132.

Tarn W-Y, Steitz JA. A novel spliceosome containing U11, U12 and U5 snRNPs excises a monor class (AT-AC) intron in vitro. Cell 1996a; 84: 801-811.

Tarn W-Y, Steitz JA. Highly diverged U4 and U6 small nuclear RNAs required for splicing rare AT-AC introns. Science 1996b; 273: 1824-1832.

Tarun SZ, Jr. and Sachs AB. A common function for mRNA 5' and 3' ends in translation initiation in yeast. Genes Dev 1995; 9:2997-3007.

Tarun SZ, Jr., Sachs AB. Association of the yeast poly(A) tail binding protein with translation initiation factor eIF-4G. Embo J 1996; 15, 7168-7177.

Tarun SZ, Jr., Wells SE, Deardorff JA and Sachs AB. 1997 Translation initiation factor eIF4G mediates in vitro poly(A) tail-dependent translation. Proc Natl Acad Sci USA 1997; 94:9046-9051.

Tay J, Hodgman R, Richer JD. The control of cyclin B1 mRNA translation during mouse oocyte maturation. Dev Biol 2000; 221:1-9.

Teerink H, Voorma HO, Thomas AA. The human insulin-like growth factor II leader 1 contains an internal ribosomal entry site. Biochim Biophys Acta 1995;1264; 403-408.

Temin HM. RNA-directed DNA synthesis. Sci Am 1972; 226:24-31.

Thanaraj TA, Argos P. Protein secondary structural types are differentially coded on messenger RNA. Protein Sci 1996; 5:1973-1983.

Tharun S, He W, Mayes AE, Lennertz P, Beggs JD, Parker R. Yeast Sm-like proteins function in mRNA decapping and decay. Nature 2000; 404:515-518.

Theimer CA, Wang Y, Hoffman DW, Krisch HM, Giedroc DP. Non-nearest neighbor effects on the thermodynamics of unfolding of a model mRNA pseudoknot. J Mol Biol 1998; 279:545-564.

Theodorakis NG, Cleveland DW . Translationally coupled degradation of mRNA in eukaryotes. In: Hershey J, Mathews M, Sonenberg N (eds) Translational control Cold Spring Harbor Laboratory Press, Cold Spring Harbor, New York 1996:631-652.

Theurkauf WE, Hazelrigg TI. In vivo analyses of cytoplasmic transport and cytoskeletal organization during *Drosophila* oogenesis: characterization of a multi-step anterior localization pathway. Development 1998; 125:3655-3666.

Thomas JD, Conrad RC, Blumenthal T. The C.elegans trans-spliced leader RNA is bound to Sm and has a trimethylguanosine cap. Cell 1988; 54: 533-539.

Tiwari S, Ramachandran S, Bhattacharya A, Bhattacharya S, Ramaswamy R. Predication of probable genes by Fourier analysis of genomic sequences. CABIOS 1997; 13: 263-270.

Tollervey D, Kiss T. Function and synthesis of small nucleolar RNAs. Curr Opin Biol 1997; 9:337-342.

Tollervey D, Lehtonen H, Carmo-Fonseca M, Hurt EC. The small nucleolar RNP protein NOP1 (fibrillarin) is required for pre-rRNA processing in yeast. EMBO J 1991; 10: 573-583.

Tomita K, Ueda T and Watanabe K. RNA editing in the acceptor stem of squid mitochondrial tRNA(Tyr). Nucleic Acids Res 1996, 24, 4987-4991.

Tomita N, Morishita R, Higaki J et al. Transient decrease in high blood pressure by *in vivo* transfer of antisense oligodeoxynucleotides against rat angiotensinogen. Hypertension 1995; 26:131-136.

Tormay P, Sawers A, Böck A. Role of stoichometry between mRNA, translation factor SelB and selenocysteyl-tRNA in selenoprotein synthesis. Mol Microbiol 1996; 21:1253-1259.

Touriol C, Morillon A, Gensac MC, Prats H, Prats AC. Expression of human fibroblast growth factor 2 mRNA is post-transcriptionally controlled by a unique destabilizing element present in the 3'-untranslated region between alternative polyadenylation sites. J Biol Chem. 1999; 274,21402-8.

Trifonov EN. Interfering contexts of regulatory sequence elements. CABIOS 1996; 12:423-429.

Trono D, Feinberg M.B, Baltimore D . HIV-1 Gag mutants can dominantly interfere with the replication of the wild-type virus. Cell 1989; 59: 113-120. CHECK citation spelling

Tuck MT. The formation of internal 6-methyladenine residues in eukaryotic messenger RNA. Int J Biochem 1992; 24: 379-386.

Türk C. Using the SELEX combinatorial chemistry process to find high affinity nucleic acid ligands to target molecules. Methods Mol Biol 1997; 67: 219-230.

Tycowski KT, Smith CM, Shu M-D, Steitz JA. A small nucleolar RNA required for site-specific ribose methylation of rRNA in *Xenopus*. Proc Natl Acad Sci USA 1996; 93: 14480-14485.

Tzfati Y, Fulton TB, Roy J, Blackburn EH. Template boundary in a yeast telomerase specified by RNA structure. Science 2000; 288:863-867.

Udem SA, Warner JR. Ribosomal RNA synthesis in *Saccharomyces cerevisiae*. J Biol Chem 1972; 248: 1412-1416.

Unrau PJ, Bartel DP. RNA-catalysed nucleotide synthesis. Nature 1998; 395:260-263.

Vagner S, Gensac MC, Maret A, Bayard F, Amalric F, Prats H and Prats AC. Alternative translation of human fibroblast growth factor 2 mRNA occurs by internal entry of ribosomes. Mol Cell Biol 1995; 15: 35-44.

Van Biesen T, Soderbom F, Wagner EG, Frost LS. Structural and functional analyses of the FinP antisense rerulatory system of the F conjugative plasmid. Mol Microbiol 1993; 10: 35-43.

Vanchiere JA, Bellini WJ, Moyer SA. Hypermutation of the phosphoprotein and altered mRNA editing in the hamster neurotrophic strain of measles virus. Virology 1995; 207:555-561.

Vanet A, Marsan L, Labigne A, Sagot MF. Inferring regulatory elements from a whole genome. An analysis of *Helicobacter pylori* sigma family of promoter signals. J Mol Biol 2000; 297:335-53.

Van Horn DJ, Eisenberg D, O'Brien CA, Wolin SL. *Caenorhabditis elegans* embryos contain only one major species of Ro RNP. RNA 1995; 1:293-303.

Van Steensel B, de Lange T. Control of telomere length by the human telomeric protein TRF1. Nature 1997; 385: 740-743.

Varani L, Spillanti MG, Goedert M, Varani G. Structural basis for recognition of the RNA major groove in the tau exon 10 splicing regulatory element by aminoglycoside antibiotics. Nucleic Acids Res 2000; 28:710-709.

Veldman GM, Brand RC, Klootwijk J, Planta RJ. Some characteristics of processing sites in ribosomal precursor RNA of yeast. Nucl Acid Res 1980; 8: 2907-2920.

Vellard M, Sureau J, Soret C, Martinerie C, Perbol B. A potential splicing factor is encoded by the opposite strand of the trans-spliced c-myb exon. Proc Natl Acad Sci 1992; 89: 2511-2515.

Venema J, Tollervey D. Processing of pre-ribosomal RNA in *Saccharomyces cerevisiae*. Yeast 1995; 11: 1629-1650.

Venema J, Tollervey D. Ribosome synthesis in *Saccharomyces cerevisiae*. Annu Rev Genet 1999; 33: 261-311.

Venema J, Henry Y, Tollervey D. Two distinct recognition signals define the site of endonucleolytic cleavage at the 5' end of yeast 18S rRNA. EMBO J 1995; 14 4883-4892.

Veyrune JL, Campbell GP, Wiseman J, Blachard JM, Hesketh JE. A localisation signal in the 3' untranslated region of c-myc mRNA targets c-myc mRNA and beta-globin reporter sequences to the perinuclear cytoplasm and cytoskeletal-bound polysomes. J Cell Sci 1996; 109:1185-1194.

Villsen ID, Vester B, Douthwaite S. ErmE methyltransferase recognizes features of the primary and secondary structure in a motif within domain V of 23 S rRNA. J Mol Biol 1999; 286: 365-374.

von Hippel PH , Kowalczykowski SC, Lonberg N, Newport JW, Paul LS, Stormo GD, Gold L. Autoregulation of gene expression. Quantitative evaluation of the expression and function of the bacteriophage T4 gene 32 protein system. J Mol Biol 1982; 162: 795-818.

Wagner EGH, Simons RW. Antisense RNA control in bacteria, phages, and plasmids. Annu Rev Microbiol 1994; 48: 713-742.

Walter NG, Yang N, Burke JM. Probing non-selective cation binding in the hairpin ribozyme with Tb(III). J Mol Biol 2000; 298:539-55.

Warnecke JM, Furtse JP, Hardt WD, Erdmann VA, Hartmann RK. Ribonuclease P (RNaseP) RNA is converted to a Cd^{++}- ribozyme by a single Rp-phosphorothioate modification in the precursor tRNA at the RNAse P cleavage site. Proc Natl Acad Sci USA 1996; 93: 8924-8928.

Wassarmann DA, Steitz JA. Interactions of small nuclear RNAs with precursor messenger RNA during in vitro splicing. Science 1992; 257: 1918-1925.

Wasserman WW, Palumbo M, Thompson W, Fickett JW, Lawrence CE. Human-mouse genome comparisons to locate regulatory sites. Nat Genet 2000; 26:225-228.

Watson JD. Involvement of RNA in the synthesis of proteins. Science 1963; 140:17-26.

Wei J, Theil EC. Identification and characterization of the iron regulatory element in the ferritin gene of a plant (soybean). J Biol Chem 2000; 275:17488-17493.

Wei P, Garber ME, Fang SM, Fischer WH, Jones KA. A novel CDK9-associated C-type cyclin interacts directly with HIV-1 Tat and mediates its high-affinity, loop-specific binding to TAR RNA. Cell 1998; 92:451-462.

Weiss G, Houston T, Kastner S, Johrer K, Grunewald K, Brock JH. Regulation of cellular iron metabolism by erythropoietin: activation of iron-regulatory protein and upregulation of transferrin receptor expressin in erythroid cells. Blood 1997; 89:680-687.

Wells SE, Hillner PE, Vale RD, Sachs AB. Circularization of mRNA by eukaryotic translation initiation factors. Mol Cell 1998; 2, 135-140.

Werner M, Feller A, Messenguy F, Piérard A. The leader peptide of yeast gene CPA1 is essential for the translational repression of its expression. Cell 1987; 49: 805-813.

Werstuck G, Green MR. Controlling gene expression in living cells through small molecule-RNA interactions. Science 1998; 282:296-298.

Wevrick A, Kerns JA, Francke U. Identification of a novel paternally expressed gene in the Prader-Willi syndrome region. Hum Mol Genet 1994; 3: 1877-1882.

Wharton RP, Struhl G. RNA regulatory elements mediate control of *Drosophila* body pattern by the posterior morphogen nanos. Cell 1991; 67:955-967.

Wickens M. In the beginnning is the end: regulation of poly(A) addition and removal during early development. TIBS 1990; 15:320-323 (check end page).

Wickens M, Anderson P, Jackson RJ. Life and death in the cytoplasm: messages from the 3'end. Curr Opin Genet Dev 1997; 7:220-232.

Wickens M, Goodwin EB; Kimble J, Strickland S, Hentze MW . Translational control of developmental decisions. In: Sonenberg N, Hershey JWB, Mathews MB (eds) Translational control of gene expression Cold Spring Harbor Laboratory Press, Cold Spring Harbor, New York 2000:295-370.

Wickens M, Takayama K. Deviants-or emissaries. Nature 1994; 367: 17-18.

Wightmann B, Ha I, Ruvkun G. Posttranscriptional regulation of the heterochronic gene lin-14 by lin-4 mediates temporal pattern formation in *C. elegans*. Cell 1993; 75: 855-862.

Wightman B, Burglin TR, Gatto J, Arasu P, Ruvkun G. Negative regulatory sequences in the lin-14 3' -untranslated regions are necessary to generate a temporal switch during *Caenorhabditis elegans* development. Genes Dev 1991; 5: 1813-1824

Will CL, Schneider C, Reed R, Luhrmann R. Identification of both shared and distinct proteins in the major and minor spliceosomes. Science 1999; 284:2003-2005.

Willard HF, Salz HK. Remodelling chromatin with RNA. Nature 1997; 386: 228-229.

Williams DJ, Hall KB. Experimental and computational studies of the G[UUCG]C RNA tetraloop. J Mol Biol 2000; 297:1045-1061.

Williams KP, Ciafre S, Tocchini-Valentini GP. Selection of novel Mg^{++} dependent self cleaving ribozymes. EMBO J. 1995; 14, 4551-4557.

Wilson JE, Pestova TV, Hellen CU, Sarnow P. Initiation of protein synthesis from the A site of the ribosome. Cell 2000a; 102:511-520.

Wilson JE, Powell MJ, Hoover SE and Sarnow P. Naturally occurring dicistronic cricket paralysis virus RNA is regulated by two internal ribosome entry sites. Mol Cell Biol 2000b; 20:4990-4999.

Wilson KS, Conant CR, von Hippel PH. Determinants of the stability of transcription elongation complexes: interactions of the nascent RNA with the DNA template and the RNA polymerase. J Mol Biol 1999; 289:1179-1194.

Winzeler EA, Richards DR, Conway AR, Goldstein AL, Kalman S, McCullough MJ, McCusker JH, Stevens DA, Wodicka L, Lockhart DJ, Davis RW. Direct allelic variation scanning of the yeast genome. Science 1998; 281:1194-1197.

Wissinger B, Schuster W, Brennicke A. Trans splicing in *Oenothera* mitochondria: nad1 mRNAs are edited in exon and *trans*-splicing group II intron sequences. Cell 1991; 65: 473-482.

Wittop-Koning TH, Schümperli D. RNAs and ribonucleoproteins in recognition and catalysis. Eur J Biochem. 1994; 219: 25-42.

Wolfertstetter F, Frech K, Herrmann G, Werner T. Identification of functional elements in unaligned nucleic acid sequences by a novel tuple search. CABIO 1997; 12: 71-80.

Wu Q, Krainer AR. U1-mediated exon definition interactions between AT-AC and GT-AG introns. Science 1996; 274: 1005-1008.

Wu S, Romfo CM, Nilsen TW, Green MR. Functional recognition of the 3' splice site AG by the splicing factor U2AF35. Nature 1999; 402:832-835.

Wu TH, Liao SM, McLure WR, Susskind MM. Control of gene expression in bacteriophage P22 by a small antisense RNA. II Characterization of mutatnts defectice in repression. Genes Dev. 1987; 1:204-212.

Yamanaka K, Minato N, Iwai K. Stabilization of iron regulatory protein 2, IRP2, by aluminium. FEBS Lett 1999; 462:216-220.

Yamanaka S, Poksay KS, Arnold KS, Innerarity TL. A novel repressor mRNA is edited extensively in livers containing tumors caused by the transgene expression of the apoB mRNA-editing enzyme. Genes Dev 1997; 11: 321-333.

Yanofsky C, Konan KV, Sorsero JB. Some novel transcription attenuation mechanisms usd by bacteria. Biochimie 1996;78:1017-1024.

Yarnell WS, Roberts JW. Mechanism of intrinsic transcription termination and antitermination. Science 1999; 284:611-615.

Ye X, Fong P, Iizuka N, Choate D and Cavener DR. Ultrabithorax and Antennapedia 5' untranslated regions promote developmentally regulated internal translation initiation. Mol Cell Biol 1997; 17, 1714-1721.

Yean SL, Lin RJ. U4 small nuclear RNA disassociates from a yeast spliceosome and does not participate in the subsequent splicing reaction. Mol Cell Biol 1991; 11: 5571-5577.

Yokobori SI, Paabo S. tRNA editing in metazoans. Nature 1995; 377:490.

Yong TJ, Gan YY, Toh BH, Sentry JW. Human CKIalpha(I) and CKIalpha(S) are encoded by both 2.4- and 4.2.-kb transcripts, the longer containing multiple RNA-destabilizing elements.Biochim Biophys Acta 2000;1492:425-433.

Young LS, Dunstan HM, Witte PR, Smith TP, Ottonello S, Spragu KU. A class II transcription factor composed of RNA. Science 1991; 252:542-546.

Yu Y-T, Steitz JA. Site-specific crosslinking of mammalian U11 and U6atac to the 5' splice site of an AT-AC intron. Proc Natl Acad Sci USA 1997; 94: 6030-6035

Yu Y-T, Scharl EC, Smith CM, and Steitz JA The growing world of small nuclear ribonucleoproteins. In: Gesteland RF, Cech TR, and Atkins JF (eds) The RNA World, 2nd edn 1999:487-524.

Zachar Z, Chou TB, Bingham PM. Evidence that a regulatory gene autoregulats splicing of its own transcript. EMBO J 1987; 6:4105-4111.

Zacharias M, Sklenar H. Conformational analysis of single-base bulges in A-form DNA and RNA using a hierarchical approach and energetic evaluation with a continuum solvent model. J Mol Biol 1999; 289:261-275.

Zakian VA. Telomers: beginning to understand the end. Science 1995; 270: 1601-1607.

Zahringer J, Baliga BS, Munro HN. Proc Natl Acad Sci USA 1976; 73:857-861.

Zamore PD, Tuschl T, Sharp PA, Bartel DP. RNAi: double stranded RNA directs the ATP-dependent cleavage of mRNA at 21 to 23 nucleotide intervals. Cell 2000; 101:25-33.

Zavanelli MI, Ares M. Efficient association of U2 snRNPs with pre-mRNA requires an essential U2 RNA structural element. Genes Dev 1991; 5: 2521-2533.

Zeffman A, Hassard S, Varani G, Lever A. The major HIV-1 packaging signal is an extended bulged loop whose structure is altered on interaction with the Gag polyprotein. J Mol Biol 2000; 297:877-893.

Zhang M, Pierce RA, Wachi H, Mecham RP, Parks WC. An open reading frame element mediates posttranscriptional regulation of tropoelastin and responsivness to transforming growth factor beta1. Mol Cell Biol 1999; 19:7314-7326.

Zheng H, Fu TB, Lazinski D, Taylor J. Editing on the genomic RNA of human hepatitis delta virus. J Virol 1992; 66: 4693-4697.

Zimmerly S, Moran JV, Perlman PS, Lambowitz AM. Group II intron reverse transcriptase in yeast mitochondria. Stabilization and regulation of reverse transcriptase activity by the intron RNA. J Mol Biol 1999; 289:473-490.

Zorio DA, Blumenthal T. Both subunits of U2AF recognize the 3' splice site in Caenorhabditis elegans. Nature 1999; 402:835-838.

Zuker M. On finding all suboptimal foldings of an RNA molecule. Science 1989; 244: 48-52.

Zuker M, Mathews DH, Turner DH. Algorithms and thermodynamics for RNA secondary structure prediction: a practical guide in RNA biochemistry and biotechnology. NATO ASI Series, Kluwer, Dordrecht, 1999: 11-43.

10. Subject Index

A

aauaaa element 19, 20, 36-37, 85, 96, 104, 163-165, 205
affinity purification 72, 76, 121, 216
aleatoric library 83
antisense RNA 7-8, 28, 32, 39, 126, 129, 135-136, 166, 174, 181, 183, 192-193, 195, 197, 200, 203, 211, 221-222
apoptosis 20, 128, 134, 202
AUG codon 25, 35, 74, 108, 160, 202, 211, 214
autoregulatory RNA 27

B

bandshift assays 71, 95
bicoid (bcd) 24, 38, 70, 118, 165-166, 197, 208

C

C. elegans (Caenorhabditis elegans) 21, 28, 112-113, 135, 144-145, 164, 167, 172, 174, 187, 199, 200, 203, 205, 211, 219, 222
cap, guanosine 17, 37, 120, 140, 158, 214, 216-217, 220
catalytic RNA X, 5-6, 8, 18, 30, 70, 72, 91, 107, 114-115, 123, 131, 168-169, 192, 208, 219
       hairpin motif 6, 25, 30, 33, 38, 78, 80, 94-95, 113, 122, 131, 133, 139, 164, 170, 192, 194, 197, 206,   209, 211, 216, 221
      hammerhead motif 6, 8, 30, 39, 80-81, 131-132, 136, 201, 216
      ribozymes 2, 8, 18, 71, 79, 80, 81, 114, 123, 126, 130, 131, 134, 196, 203, 205-206, 209, 217, 219, 222
CBC (cap-binding complex) 140
cDNA 78, 160-161, 164, 171, 192, 205
CPE (cytoplasmic polyadenylation element) 25, 36, 163-165, 209
cross-linking 69, 76, 91, 168, 215
crystallization 79, 196-197
cytoplasmic polyadenylation 25, 36, 163-165, 200

D

differential splicing 124, 140, 178

V

viral RNA IX, X, 1, 6, 19, 30-32, 35, 38, 126, 134, 147, 153, 162, 170, 218-219

X

Xist RNA 33
xol-1 172

Y

Y-RNA 29

Printing (Computer to Film): Saladruck, Berlin
Binding: Stürtz AG, Würzburg